John Ernst

OPTICAL
SENSING TECHNIQUES
AND
SIGNAL PROCESSING

OPTICAL SENSING TECHNIQUES
AND
SIGNAL PROCESSING

Tudor E. Jenkins

Department of Physics,
University College of North Wales,
Aberystwyth

Prentice/Hall International

Englewood Cliffs, NJ London Mexico New Delhi
Rio de Janeiro Singapore Sydney Tokyo Toronto

Library of Congress Cataloging-in-Publication Data

Jenkins, Tudor E.
  Optical sensing techniques and signal processing.

  Bibliography: p.
  Includes index.
  1. Optical detectors. 2. Optical measurements.
3. Signal processing. I. Title.
TA1632.J46   1987   621.3815'42   86-30661
ISBN 0-13-638107-3

British Library Cataloguing in Publication Data

Jenkins, Tudor E.
  Optical sensing techniques and signal
  processing.
  1. Signal processing 2. Optical data
  processing
  I. Title
  621.36'7   TK5102.5

  ISBN 0-13-638107-3

Prentice-Hall Inc., Englewood Cliffs, New Jersey
Prentice-Hall International (UK) Ltd, London
Prentice-Hall of Australia Pty Ltd, Sydney
Prentice-Hall Canada Inc., Toronto
Prentice-Hall Hispanoamericana S.A., Mexico
Prentice-Hall of India Private Ltd, New Delhi
Prentice-Hall of Japan Inc., Tokyo
Prentice-Hall of Southeast Asia Pte Ltd, Singapore
Editora Prentice-Hall do Brasil Ltda, Rio de Janeiro

Printed and bound in Great Britain for Prentice-Hall International
(UK) Ltd, 66 Wood Lane End, Hemel Hempstead, Hertfordshire,
HP2 4RG by A. Wheaton & Co. Ltd, Exeter.

1 2 3 4 5 91 90 89 88 87

ISBN 0-13-638107-3

To my wife, Susan, and my children, Bethan and Morgan,
without whom this book would have been finished much sooner.

Contents

Preface

This book has been written as the result of experiences in several university physics departments over many years with students undertaking final year experimental projects or postgraduate students beginning their research careers. For probably the first time, the student is faced with the prospect of devising and specifying his own experimental system. While the information needed to complete the task is undoubtedly available in the literature, it is widely dispersed and considerable patience is needed to track it down. It is the aim of this book to collect such information. It is not intended as a rigorous mathematical treatment of the subject but to indicate the problems and sources of error which can occur in designing a practical measurement system and also to indicate the major signal processing techniques which may be employed. It is then hoped that the designer may determine the experimental procedures best suited to his needs and, if necessary, gain deeper insight by following up the references provided.

The level assumed by the book is that of final year undergraduate or postgraduate physics student. The level of mathematics needed is not particularly high but a basic knowledge of semiconductor physics is assumed for the chapter on photodetectors. I have also assumed a knowledge of microcomputers and programming. Judging by the many students I encounter, this is not an unreasonable assumption.

The first chapter reviews the formalism needed to describe signals and noise and looks at the physical origins of noise. Chapter 2 discusses optical detectors as an important example of transducer. The factors needed to specify the detector – its spectral responsivity, its speed of response, and its ability to detect weak signals in the presence of noise – are presented. The signal-to-noise ratios for several types of photodetector are determined in detail. Chapter 3 contains a discussion of photodetectors in linear or area arrays. In Chapter 4, we encounter signal preprocessing; in particular, amplification and filtering of signals. The operational amplifier is discussed in detail, emphasizing the errors which can be introduced by the use of such amplifiers. Chapter 5 introduces some of the major signal processing techniques – photon counting, phase sensitive detection, signal averaging and heterodyne detection. Chapter 6 bows to the inevitable dominance of the microprocessor in the field of signal processing. The chapter does not seek to describe in any detail the microprocessor itself (the book market is saturated with books on microprocessors) but rather addresses the problems and errors associated with digitizing a signal and the ways in which the digitized and stored signals may be processed. Signal averaging, digital filtering, the fast Fourier transform, and correlation are all discussed.

There are a number of people that I must thank for help and support throughout the preparation of this book. I should like to thank Prof. R.A. Stradling and Mr. F.C. Evans of St. Andrews University for early encouragement of my work and Prof. L. Thomas of the University College of Wales, Aberystwyth, for continued support of my work. I should also like to thank Mrs. M. Trethewey and Mrs M. Garnett for their typing of much of the script and the creator of the Wordstar program which allowed me to type the rest. My parents must also bear some responsibility for this work, since, without their perseverance, I should now almost certainly be hewing coal in some Welsh pit. Finally, my wife and children have had to suffer a lot of my anguish during the writing and to them I give my heartfelt thanks.

1

Noise

1.1 INTRODUCTION

One of the most basic activities in science and engineering is the measurement of physical quantities such as length, temperature, or light intensity. In any real situation it is impossible to determine the true value of the measureand but the experimenter should be able to indicate the range of values within which the true value occurs. Such errors can be broadly characterized as systematic errors and random errors. A common example of systematic error is the so-called mains pick-up which results in a 50 Hz sinusoidal voltage superimposed on a voltage which is being measured. Such systematic errors can be minimized by careful design of the experiment (for a discussion of these errors see Cunningham 1981).

Random errors are fundamental to the corpuscular nature of matter and are caused by spontaneous fluctuations in certain physical quantities. They are random in the sense that their values from moment to moment are different and that the amplitude of a random error can only be predicted or calculated in statistical terms. Such random errors will set the limit to the accuracy and sensitivity of any measurement (Fellgett and Usher 1980).

Any instrument designed to measure a physical quantity will consist of a number of well defined components. The 'front end' of such a system is usually a transducer, which will convert the physical quantity into an electrical voltage or current. The list of transducers is enormous, but probably the most important and widely used are those which convert optical signals into electrical ones, such as photomultipliers, photodiodes, and bolometers. As a result of the basic fluctuations in the signal, the output of even a perfect transducer will consist of the 'wanted' signal plus a randomly fluctuating component which is generally known as 'noise'. Unfortunately, transducers are not perfect and will themselves add noise to the signal. In addition, because of the nonideal nature of materials used in transducers (for example, the nonunity quantum efficiency of photodetectors), transducers will limit the sensitivity of the instrument. The signal output from the transducer is then subject to a signal conditioning process. This usually consists of amplification to increase the basic signal level and filtering to remove unwanted noise components. If the ratio of signal to noise is still not great enough, signal processing techniques must be used, and both analog and digital techniques may be employed. It is the object of this book

1

to investigate the different areas of the measurement instrument, to determine the fundamental limits in the sensitivity and accuracy of the components of the instrument.

1.2 CHARACTERIZATION OF NOISE

There are several parameters which are used to characterize noise. The first of these is the mean. If there are n samples of the random variable $x_1, x_2, \ldots, x_n$, the mean value will be

$$\langle x \rangle = \frac{x_1 + x_2 + \ldots + x_n}{n}$$

In addition to this mean, the scatter of the random quantities around the mean is important. The most common way of providing this information is by the standard deviation σ which is defined as the square root of the average of the sum of the squares of the deviations of the variables x_j from the mean:

$$\sigma = \left(\frac{1}{n} \sum_{j=1}^{n} (x_j - \langle x \rangle)^2 \right)^{1/2} = \frac{(x_1 - \langle x \rangle)^2 + (x_2 - \langle x \rangle)^2 + \ldots + (x_j - \langle x \rangle)}{n}$$

σ^2 is sometimes known as the variance of the distribution. For example, if we had sample voltages 5, 3, 7, 2, 8, 5 V, then

$$\langle x \rangle = \frac{5 + 3 + 7 + 2 + 8 + 5}{6} = 5 \text{ V}$$

$$\sigma = \left(\frac{(5-5)^2 + (3-5)^2 + (7-5)^2 + (2-5)^2 + (8-5)^2 + (5-5)^2}{6} \right)^{1/2}$$

$$= 2.45 \text{ V}$$

More information about the scatter of values can be obtained from the probability density function, which provides the probability that a signal lies within a specified range of values. Suppose a signal is made up of voltages between 2 and 12 V. This range can be divided up into, say, ten intervals and a histogram can be plotted giving the probability that the voltage lies within a given 1 V range (see Fig. 1.1). The number of intervals may be increased and the histogram will approach a smooth curve as the interval width approaches zero. This curve is known as the probability density function, and the area under this curve between any two voltages will give the probability that the voltage will be found between these two values.

A particularly important distribution function in a discussion of noise is the Gaussian (or normal) distribution. Suppose an observer is monitoring the noise at the output of an amplifier. This noise is due to a large number of random phenomena, each of which has its own statistical distribution. However, a theorem in probability theory (the so-called Central Limit

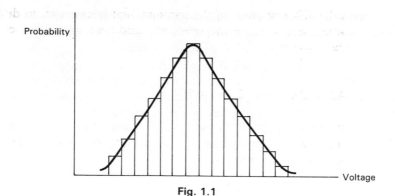

Fig. 1.1

Theorem) tells us that, when many random variables are employed, the overall distribution will obey Gaussian statistics. The probability density function for this distribution is

$$p(x) = \frac{1}{\sigma\sqrt{2\pi}} \exp\left[\frac{-x^2}{2\sigma^2}\right]$$

and is shown in Fig. 1.2. Marked on this figure are the values of x: $\pm\sigma$, $\pm2\sigma$, $\pm3\sigma$. It can be seen that $p(x)\,dx$ becomes very small as x/σ becomes very large. The probability that a Gaussian random variable, such as random noise, will be within $\pm\sigma$ of its mean value is 0.68 (i.e. 68 percent) while the probability of being within $\pm2\sigma$ is 0.95 and within $\pm3\sigma$ is 0.99.

An important property of Gaussian distributions is what happens when several such distributions are added together. This is particularly important

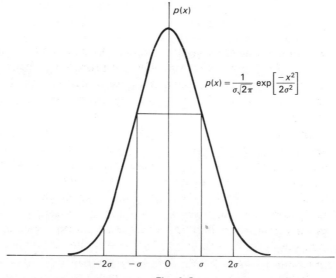

Fig. 1.2

in the area of signal averaging, where this property is used to increase the signal-to-noise ratio of a waveform. If n Gaussian distributed waveforms are added together, the resulting distribution is also Gaussian with a mean value given by the sum of the individual means and a variance given by the sum of the variances. Let us suppose now that we have a voltage which consists of a steady signal value plus a random noise contribution with standard deviation σ. The voltage is divided up into n slices and the mean value and standard deviation measured and recorded for each slice. These values are added together and the average value over the n slices computed for the mean and standard deviation.

The total mean voltage is:

$$\langle v \rangle = \langle v_1 \rangle + \langle v_2 \rangle + \ldots + \langle v_n \rangle$$

$$= n \langle v \rangle$$

so that the average of the mean is simply $\langle v \rangle$.

The total variance is

$$\sigma^2 = \sigma_1{}^2 + \sigma_2{}^2 + \ldots + \sigma_n{}^2$$

$$= n\sigma^2$$

so that the total standard deviation is

$$\sigma = \sigma\sqrt{n}$$

and the average standard deviation is

$$\sigma_{av} = \frac{\sigma\sqrt{n}}{n} = \frac{\sigma}{\sqrt{n}}$$

Thus, as we increase the number of distributions we add together, the mean value remains the same but the average standard deviation will decrease. It follows therefore that the signal-to-noise ratio after averaging will increase. For example, suppose the average value of the signal was 2 V and the standard deviation 0.5 V. Without averaging, the signal-to-noise ratio is $2/0.5 = 4$. After averaging nine times, the average standard deviation is $0.5/\sqrt{9}$ and the signal-to-noise ratio is $2\sqrt{9}/0.5 = 12$, an increase of three times.

The discussion so far has been concerned with manipulation of instantaneous values of signals and noise, but has not considered how rapidly such waveforms are varying. Consider the two waveforms shown in Fig. 1.3.

The waveforms may well have the same mean and the same standard deviation but obviously vary at different rates. It is necessary to study in some detail the relationship between the time response of a waveform and its frequency content.

The starting point for such a discussion is the Fourier series. Any periodic function of time with period T can be decomposed into a series of sinusoidal

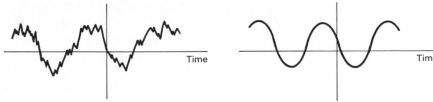

Fig. 1.3

components having frequencies which are harmonics of the fundamental frequency $f = 2\pi/T$. A simple example of this is the square wave having an average value of zero. In terms of its Fourier components, this can be written

$$f(t) = \sin \omega t + \tfrac{1}{3} \sin 3 \, \omega t + \tfrac{1}{5} \sin 5 \, \omega t \, \ldots$$

$$= \sum_{n=0}^{\infty} \frac{(-1)^n}{(2n+1)} \sin (2n+1) \, \omega t$$

where $\omega = \dfrac{2\pi}{T}$.

Figure 1.4 shows that a reasonable square wave can be constructed from the first few Fourier components.

It is more convenient to express the periodic function in terms of complex

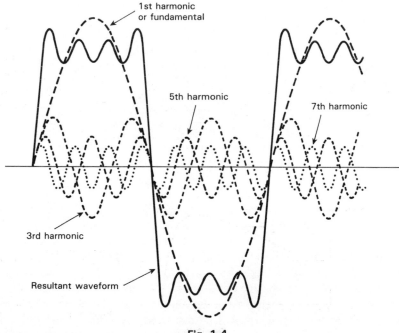

1st harmonic or fundamental

5th harmonic

7th harmonic

3rd harmonic

Resultant waveform

Fig. 1.4

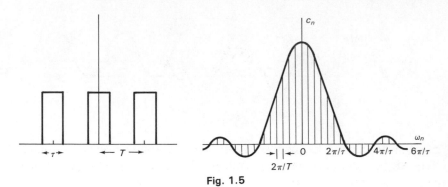

Fig. 1.5

Fourier coefficients

$$f(t) = \sum_{n=-\infty}^{+\infty} c_n e^{j\omega_n t}$$

where c_n is a complex quantity and so contains information on both amplitude and phase of components of $f(t)$ of frequency ω_n.

An important example of such an analysis is that of a series of rectangular pulses, shown in Fig. 1.5, together with its frequency spectrum (a plot of c_n against ω_n).

(Note that we are able to plot c_n against ω_n on one graph only because the phase of the components take only two values. 0 (positive values of c_n) and π (negative values of c_n). Two things can be obtained from this frequency response. Firstly, consider what happens as T changes. For a fixed pulse width τ, as T increases, the frequency components crowd more closely together so that there are more low frequency components, while as T decreases, the number of high frequency components increases. Thus, we can associate rapidly varying signals with high frequency Fourier components. Furthermore, as $T \to \infty$, the line spectrum approaches a smooth frequency spectrum. Secondly, as ω_n increases, the amplitude of the higher frequency components decreases rapidly and most of the energy in the pulses is confined in the frequency range $0 < \omega_n < 2\pi/\tau$. As an approximation, we can ignore those frequency components outside this range, and talk of the bandwidth of the signal as

$$\Delta f = \frac{1}{\tau}$$

This concept of bandwidth applies also to instrumentation, and in this case refers to the range of frequencies which the instrument can handle. In general, an instrument will have a different bandwidth from the signal and when the signal passes through the instrument, the output frequency spectrum will depend of the relative bandwidth of signal and instrument. If the bandwidth of the instrument is smaller than that of the signal, the signal at the output of the instrument will be distorted, i.e. some of its frequency components will be missing. We will return to this point in later chapters.

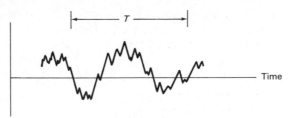

Fig. 1.6

In real life, we are faced with nonperiodic functions and the arguments of the Fourier series are extended to these functions by the introduction of the Fourier integral.

The nonperiodic function is first observed for some time T (Fig. 1.6). Within the time T, the waveform can be expanded as a Fourier series with fundamental period T. The segment of the waveform is made to repeat itself artificially outside the time interval T. As $T \to \infty$, the resultant series becomes the Fourier integral, which, for a non-periodic function $f(t)$ is

$$f(t) = \frac{1}{2\pi} \int_{-\infty}^{+\infty} F(\omega)\, e^{j\omega t}\, d\omega$$

where

$$F(\omega) = \int_{-\infty}^{+\infty} f(t)e^{-j\omega t}\, dt$$

The result of such an analysis on a single pulse can be readily seen by modifying the Fourier series result on the sequence of pulses, letting $T \to \infty$. The frequency spectrum will then become a continuous one, which is known as the spectral density. The bandwidth of such a pulse is still

$$\Delta f = 1/\tau$$

The equation:

$$F(\omega) = \int_{-\infty}^{+\infty} f(t)\, e^{-j\omega t}\, dt$$

is known as the Fourier transform of the function $f(t)$. $f(t)$ and $F(\omega)$ are known as a Fourier transform pair. The Fourier transform is a powerful tool in the analysis of linear systems and it is used extensively in digital signal processing. As an example, we can compute the Fourier transform of

$$f(t) = \cos \omega_2 t$$

Then it follows that

$$F(\omega) = \int_{-\infty}^{+\infty} \cos \omega_1 t\ e^{-j\omega t}\, dt$$

$$= \int_{-\infty}^{+\infty} \frac{1}{2}\, [e^{j\omega_1 t} + e^{-j\omega_1 t}]e^{-j\omega t}\, dt$$

$$= \frac{1}{2}\, [\delta(\omega_1 + \omega) + \delta(\omega_1 - \omega)]$$

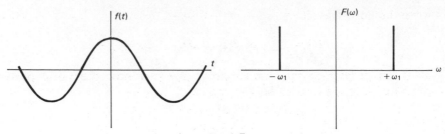

Fig. 1.7

where
$$\delta(x) = 1 \text{ if } x = 0$$
$$\delta(x) = 0 \text{ if } x \neq 0$$

See Fig. 1.7.

There are several useful properties of the Fourier transform that we will need later. They are easily proved but will simply be stated here (see, for example, Bracewell 1965, for details). Three properties of particular significance are:

1. If we add two time functions together, then the Fourier transform of the sum is the sum of the Fourier transforms of the individual functions

$$\mathscr{F}[f(t) + g(t)] = F(\omega) + G(\omega)$$

where $\mathscr{F}$ represents the Fourier transform operation.

2. The Fourier transform of the product of two time functions is the product of the Fourier transforms of the individual time functions

$$\mathscr{F}[f(t) \cdot g(t)] = F(\omega) \cdot G(\omega)$$

This is known as the convolution theorem and is of great importance in the area of digital filtering.

3. The similarity theorem states

$$\mathscr{F}[f(at)] = \frac{1}{a} F\left(\frac{\omega}{a}\right)$$

There are difficulties in applying the above concepts to random noise, since the noise function is not analytic. The ideas can be applied to a specific segment of noise, but the frequency spectrum will be different from that of any other segment of noise, so that the result from any one segment will not provide a spectrum typical of the frequency content of the random signal. Instead, the concept of the power density spectrum is used to characterize the frequency content of the random signal. This describes the power of the different frequency components in the noise but not the phase, since the phase of any signal can be defined if the origin in time can be defined; there is no way of selecting such an origin for random processes.

Consider a segment of noise voltage observed for T seconds. We can write the noise waveform

$$v(t) = \sum_{n=-\infty}^{+\infty} a_n \exp(i\omega_n t), \qquad \omega_n = \frac{2\pi n}{T}$$

a_n is, in general, a random variable. Consider now the noise power generated in a resistor R. This is

$$P = \frac{\langle v(t)^2\rangle}{2R} = \frac{1}{T}\int_0^T \frac{\langle v(t)^2\rangle}{2R}\,dt = \sum_{n=-\infty}^{+\infty} \frac{|a_n{}^2|}{2R}$$

The power in one of the Fourier components is $|a_n{}^2|^2/2R$ and is the power contained in the frequency band between the component at a frequency ω_{n-1} and that at ω_{n+1}. If we write this frequency band as Δf, then we can write the power density $G(f)$ at the particular component frequency as

$$\text{Power density } G(f) = \frac{|a_n|^2}{2R\,\Delta f}$$

or

$$\frac{|a_n|^2}{2R} = G(f)\,\Delta f$$

The total power in all the frequency components contained in the noise voltage is

$$P = \sum_{n=-\infty}^{+\infty} \frac{|a_n|^2}{2R} = \sum_{n=-\infty}^{+\infty} G(f)\,\Delta f$$

If we now let $T \to \infty$ so that $\Delta f \to 0$, the total power contained in the noise is

$$P = \int_{-\infty}^{+\infty} G(f)\,df$$

In practice, the noise will be bandwidth limited and the limits in the integral can be replaced by $\pm\Delta f$.

A commonly occurring power density spectrum is one where the power density is uniform over all frequencies, i.e.:

$$G(f) = N_0, \text{ a constant.}$$

This is known as white noise. If this noise is bandwidth limited, then

$$P = \int_{-\Delta f}^{+\Delta f} N_0\,df = 2N_0\,\Delta f$$

It is usually more convenient to use the spectral density of the voltage or of the current rather than the spectral density of the power. In analogy with the power spectral density, these are defined as

$$\langle v^2(t)\rangle = \int_{-\infty}^{+\infty} G_v(f)\,df \qquad \text{(voltage spectral density)}$$

$$\langle i^2(t)\rangle = \int_{-\infty}^{+\infty} G_i(f)\,df \qquad \text{(current spectral density)}$$

The units are V^2/Hz for the voltage spectral density and A^2/Hz for the current spectral density. It is easy to see that

$$|G(f)|^2 = G_v(f) \cdot G_i(f)$$

An alternative method of specifying the frequency content of a signal is through the autocorrelation function. This function measures how strongly a signal $x(t)$ at different times are correlated. If the value of signal at time t_1 is $x(t_1)$, the autocorrelation function will indicate how the signal at an other time t_2 $x(t_2)$ can be predicted knowing $x(t_1)$. Mathematically, the autocorrelation function of $x(t)$ is

$$R_x(t) = \langle x(t) \cdot x(t+\tau) \rangle = \lim_{T \to \infty} \frac{1}{2T} \int_{-T}^{+T} x(t)x(t+\tau) \, dt$$

Thus, $R_x(\tau)$ is the average value of $x(t)$ with the same signal shifted backward in time by an amount τ. $R_x(\tau)$ is plotted for all values of τ.

The autocorrelation function for a simple periodic waveform is easily obtained graphically by dividing the signal into a number of equal time intervals and performing the calculation

$$R_x(\tau) = \frac{1}{N} \sum_{j=1}^{N} x(j\delta)x(j\delta + \tau)$$

where τ is given by $\tau = n\delta$. n is allowed to vary from 0 to as large a value as is necessary.

In general, the autocorrelation function of any periodic function is another periodic function:

If
$$x(t) = \frac{a_0}{2} + \sum_{n=1}^{\infty} (a_n \cos n\omega t + b_n \sin n\omega t)$$

then
$$R_x(\tau) = \frac{a_0^2}{4} + \frac{1}{2} \sum_{n=1}^{\infty} (a_n^2 + b_n^2) \cos n\omega \tau$$

If $x(t)$ is a randomly fluctuating quantity, $R_x(\tau)$ will go to zero as τ becomes very large, since the value of x at one time will have no effect on the distribution of possible values at another time if the two times are sufficiently separated. In general, for noise of bandwidth Δf and mean square power N:

$$R_x(\tau) = Ne^{-\Delta f \cdot \tau}$$

Once the autocorrelation function of a signal is known, the power density spectrum can be obtained. $G(f)$ and $R_x(\tau)$ are related by the Wiener–Khintchine relationship

$$G(f) = \int_{-\infty}^{+\infty} R_x(\tau)e^{j\omega\tau} \, d\tau$$

These are a Fourier transform pair; when the autocorrelation function of

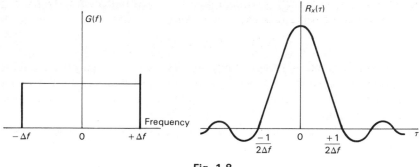

Fig. 1.8

a signal is known, the spectral density function $G(f)$ can be found by taking the Fourier transform.

For example, for band-limited white noise, the Fourier transform has the form shown in Fig. 1.8.

As the bandwidth increases, so the central peak in $R_x(\tau)$ gets narrower. Thus, the narrower the autocorrelation function the higher the frequency components present in the noise. In the limit $\Delta f \rightarrow \infty$, the autocorrelation function becomes a delta function.

1.3 THE PHYSICAL ORIGINS OF NOISE

The discussion so far has concentrated on the ways in which noise can be characterized but has neglected the problem of how to calculate from basic principles the amount of noise in a system. If we can work out the noise powers due to the various sources in a system, then the effects of the individual noise sources on the total noise output of the system can be gauged, and critical elements identified and minimized by careful circuit design. In this section, then, the basic physical origins of random noise will be investigated and the spectral density of the noise obtained. Systematic noise will not be treated since it can, in general, be more or less eliminated by careful design procedure. In all the cases treated, the random noise is caused by spontaneous fluctuations in voltage or current, which are due to the corpuscular or noncontinuous nature of matter.

1.3.1 Thermal noise

In any resistive material, such as a resistor or piece of semiconductor, free charge carriers, such as electrons or holes, in thermal equilibrium with the lattice at a temperature t will possess an average kinetic energy $\sim kT$. Thus, even in the absence of an electric field, the carriers will be moving chaotically in all directions; such moving charge carriers constitute an electric current. This current will fluctuate about a mean value of zero and will therefore lead to a fluctuating voltage across the ends of any resistance in thermal equilibrium.

An expression for this thermal noise was developed by Nyquist in 1928 using thermodynamic arguments. He showed that the mean square voltage developed across a resistor R at a temperature T K in a bandwidth Δf was

$$\langle v^2 \rangle = 4kTR \, \Delta f$$

Thus, thermal noise is white noise (at least, up to frequencies of about 10^{13} Hz where quantum mechanical effects become important); it also obeys Gaussian statistics.

At room temperature, $4kT \approx 1.6 \times 10^{-20}$ W so that for a resistor of 1 MΩ there is an r.m.s. noise voltage, in a bandwidth of 10^7 Hz, of

$$\sqrt{\langle v^2 \rangle} = \sqrt{1.6 \times 10^{-7}}$$

$$= 400 \, \mu V \text{ r.m.s.}$$

Thermal noise in a resistor can be modeled as a voltage source in series with a noiseless resistor or, using Norton's theorem, a current source in parallel with a noiseless resistor as shown in Fig. 1.9.

For the values given above, the equivalent current source has a value

$$\sqrt{\langle i^2 \rangle} = \sqrt{\frac{1.6 \times 10^{-13}}{10^6}} = 400 \text{ pA r.m.s.}$$

The spectral density of the voltage and current sources are, respectively:

$$G_v(f) = 2kTR \text{ and } G_i(f) = 2kT/R$$

taking into account positive and negative frequencies, e.g.:

$$\langle v^2 \rangle = \int_{-\Delta f}^{+\Delta f} G_v(f) \, \mathrm{d}f = 2kTR \cdot 2 \, \Delta f = 4kTR \, \Delta f$$

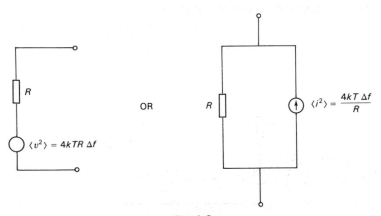

Fig. 1.9

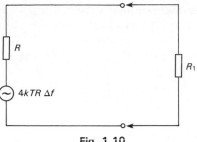

Fig. 1.10

Thus, a resistive material is a source of noise power, whose value can be calculated quite simply. Consider a noisy resistor R. It will deliver maximum power to a resistance R_1 connected across it if $R = R_1$:

$$\text{Power delivered to } R_1 = \left(\frac{\sqrt{4kTR\,\Delta f}}{R_1 + R}\right)^2 \cdot R_1$$

$$= \frac{4kTR\,\Delta f}{4R^2} \cdot R \qquad \text{if} \qquad R_1 = R$$

$$= kT\,\Delta f$$

The quantity $kT\,\Delta f$ is known as the available noise power, and has a value of 4×10^{-21} W/Hz at room temperature, independent of the value of the resistance.

The thermodynamic arguments can be illustrated by a simple example. Consider a resistor R in series with a capacitor C (Fig. 1.11).

The resistor is represented by a 'noiseless' resistor R and a noise voltage. The RC combination acts as a voltage divider and the total mean square voltage across the capacitor is

$$\langle v_c^2 \rangle = \int_{-\Delta f}^{+\Delta f} \frac{G(f)\,df}{(1 + 4\pi^2 f^2 R^2 C^2)}$$

and the energy stored in the capacitor is

$$\frac{1}{2} C\langle v_c^2 \rangle = \frac{1}{2} C \int_{-\Delta f}^{+\Delta f} \frac{G_v(f)\,df}{(1 + 4\pi^2 f^2 R^2 C^2)}$$

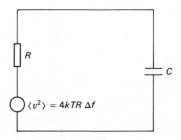

Fig. 1.11

By the equipartition theorem in thermodynamics, this has a value $\frac{1}{2} kT$:

$$\frac{1}{2} kT = \frac{1}{2} C \int_{-\Delta f}^{+\Delta f} \frac{G_v(f) \, df}{(1 + 4\pi^2 f^2 R^2 C^2)}$$

which is satisfied if

$$G_v(f) = 2kTR$$

1.3.2 Shot noise

When a current flows across a potential barrier, as, for example, across a
p–n junction or in thermionic emission from a surface, there will be a fluctuation in the number of charge carriers contributing to the current flow.
This may arise in a number of ways. For example, any carriers crossing the
barrier must have sufficient thermal energy so that the number crossing will
fluctuate in a way determined by the normal thermal fluctuations in the
position and energy distribution of those charge carriers. As we shall see in
Chapter 2, photon arrival at a detector is a random process, so that there
will be a random photogeneration of charge carriers in this case. Thus, any
current flow will exhibit random fluctuations about its mean value.

Schottky (1918) showed that the noise current is white and has a mean
square value:

$$\langle i^2 \rangle = 2eI\Delta f$$

where I is the average d.c. current flowing and e is the electronic charge.
The noise current spectral density is

$$G_i(f) = eI$$

so that

$$\langle i^2 \rangle = \int_{-\Delta f}^{-\Delta f} G_i(f) \, df = 2eI\Delta f$$

For a d.c. current of 1 mA and a bandwidth of 1 kHz

$$\sqrt{\langle i^2 \rangle} \approx 5.5 \text{ nA r.m.s.}$$

1.3.3 Flicker (1/f) noise

The noise studied so far has been of the white noise variety. However,
flicker noise, which shows itself as a fluctuation in electrical conductance,
exhibits a spectral density which is inversely proportional to the frequency.
It is, for this reason, also called $1/f$ noise and can dominate the noise in
semiconductors at sufficiently low frequency. The current spectral density

can be written

$$G_i(f) = \frac{CI^2}{f}$$

where C is a constant which depends on the type of material and its geometry. The spectral density has been measured down to 10^{-6} Hz and has been shown to be $1/f$ to high accuracy. There are problems with this formula as $f \to \infty$ where the total noise power becomes infinite (see Hooge *et al.* 1981 for a fuller discussion of this point). There is no accepted theory for $1/f$ noise and phenomena such as generation–recombination at surface states and nonohmic contacts have been postulated.

1.3.4 Generation–recombination noise

Generation–recombination noise is caused by fluctuations in the rate of thermal generation of free electrons and holes in a semiconductor. This will cause fluctuations in the carrier concentration and, hence, fluctuations in the conductivity of the semiconductor. If a current is now made to flow through the material, a fluctuating voltage will be observed across its ends.

Generation–recombination noise has a spectral density which rolls off at high frequencies

$$G(f) = \frac{G(0)}{(1 + \omega^2 \tau^2)}$$

the detailed form of the noise spectral density depends on the type of semiconductor, the level of bias, etc. These will not be reproduced here (see van der Ziel 1959). The important point is that the spectral density falls off as $1/f^2$ at high enough frequency.

1.4 NOISE POWER DUE TO SEVERAL RANDOM SOURCES

We are unlikely to be in a situation where we have a single source of noise and so must consider how we would combine several noise sources. We first look at the case where the noise voltages are produced independently and there is no relation between the instantaneous values of the voltages, that is, they are uncorrelated.

Suppose then that we have two such uncorrelated sources. The output power of these sources is simply the sum of the powers of the two sources. Thus, if we have two noise voltage sources V_1 and V_2 connected in series, then

$$\langle V^2 \rangle = \langle V_1^2 \rangle + \langle V_2^2 \rangle$$

Consider, for example, two resistors R_1 and R_2 connected in series. Then

$$\langle V^2 \rangle = 4kTR_1\,\Delta f + 4kTR_2\,\Delta f$$

$$= 4kT\,\Delta f(R_1 + R_2)$$

If we have two noise sources connected in parallel, then similarly

$$\langle I^2 \rangle = \langle I_1^2 \rangle + \langle I_2^2 \rangle$$

For our example of two resistors R_1 and R_2 connected in parallel

$$\langle I^2 \rangle = \frac{4kT\,\Delta f}{R_1} + \frac{4kT\,\Delta f}{R_2}$$

$$= 4kT\,\Delta f\left(\frac{1}{R_1} + \frac{1}{R_2}\right)$$

The equivalent noise voltage source is

$$\langle V^2 \rangle = \langle I^2 \rangle \left(\frac{1}{R_1} + \frac{1}{R_2}\right)^{-2}$$

$$= 4kT\,\Delta f\left(\frac{1}{R_1} + \frac{1}{R_2}\right)^{-1}$$

In this case, we could have combined the parallel resistance into an equivalent resistance

$$R = \left(\frac{1}{R_1} + \frac{1}{R_2}\right)^{-1}$$

and computed the noise voltage directly.

If the noise sources are partially correlated, then

$$\langle V^2 \rangle = \langle V_1^2 \rangle + \langle V_2^2 \rangle + 2C\langle V_1 \cdot V_2 \rangle$$

C is called the correlation coefficient. When $C = 0$, the noise sources are uncorrelated. When $W = \pm 1$, the sources are totally correlated.

SUMMARY

In this chapter we have considered the problems associated with measuring a 'wanted' signal in the presence of some interfering signal which we refer to as noise. In many circumstances, it is not possible to separate this wanted signal from the noise and we will need to apply some technique to improve the ratio of the signal to the noise. The first step in determining the appropriate technique to apply is a study of the characteristics of both signal and noise. The first part of this chapter was concerned with the description of a waveform. This breaks into two parts; the first is the methods of quantifying the instantaneous value of a waveform (the mean, the mean square, and the probability density of a waveform) while the second is a description of the frequency content of a waveform (the power density and the autocorrelation function).

The second part of this chapter concerned itself with the physical origins of noise, caused by the basic 'granular' or quantum nature of matter. Thermal noise, shot noise, generation–recombination noise, and flicker noise were discussed and the spectral density of these noise sources obtained. Once this has been done, we are then in a position to calculate, from first principles, the noise power in a system and to isolate critical sources of noise with a view to minimizing their effect by proper system design.

REFERENCES

Bracewell, R. 1965. *The Fourier Transform and its Applications*. New York: McGraw Hill.

Cunningham, M.J. 1981. Measurement errors and instrument inaccuracies. *Journal of Physics E: Scientific Instruments*, Vol. 13, pp. 901–908.

Fellgett P.B. and Usher, M.J., 1980. Fluctuation phenomena in instrument science. *Journal of Physics E: Scientific Instruments*, Vol. 13, pp., 1401–1406.

Nyquist, H. 1928. Thermal agitation of electric charge in conductors. *Physical Review*, Vol. 32, pp. 110–113.

Schottky, W. 1918. Spontaneous current fluctuations in various conductors. *Annalen der Physik*, Vol. 57, pp. 541–67.

Hooge, F.N., Kleinpenning, T. G. M. and Vandamme, L. K. J. 1981. Experimental studies on $1/f$ noise. *Reports on Progress in Physics (GB)*, Vol. 44, pp. 479–532.

van der Ziel, A. N. 1959. *Fluctuation Phenomena in Semiconductors*. New York: Academic Press.

PROBLEMS

1.1 Using the Gaussian probability density

$$p(x) = \frac{1}{\sqrt{2\pi\sigma^2}} \exp\left[\frac{-(x-a)^2}{2\sigma^2}\right]$$

evaluate the mean of x and its variance.

$$\int_{-\infty}^{+\infty} e^{-z^2} \, dz = \sqrt{\pi}$$

$$\int_{-\infty}^{+\infty} ze^{-z^2} \, dz = 0$$

1.2 For the Poisson distribution, show that the variance is equal to the mean value.

Photons arrive, at a photodetector, with a Poisson distribution. Show that the mean square fluctuation in the photocurrent from the detector is

$$\langle i^2 \rangle = 2e\langle I \rangle \, \Delta f$$

where e is the electronic charge, $\langle I \rangle$ the mean photocurrent and Δf the bandwidth.

1.3 A capacitor C has an initial charge Q_0. At time $t = 0$, a noiseless, ideal switch is closed and the capacitor is charged to a voltage V_F through a resistor R. Obtain an expression for the r.m.s. voltage across the capacitor when

(a) $t \ll 0.5\,RC$
(b) $t \gg 0.5\,RC$

1.4 White noise is applied to the network shown below. Calculate:

(a) the spectral density of the output noise
(b) the autocorrelation function of the output noise
(c) the average noise output.

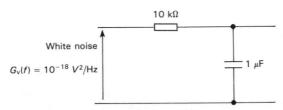

1.5 Calculate the mean square noise voltage of a 1 MΩ resistor at $T = 300$ K and $T = 100$ K.

1.6 In practice, a resistor will always be shunted by some stray capacitance C, giving a noise model of a resistor as shown in the following diagram.

By integrating the output noise spectral density over the whole frequency spectrum, show that the total noise power available from the resistor is kT/C, where k is Boltzmann's constant, and T the temperature.

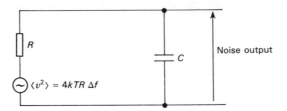

1.7 The current flowing through a p–n junction is given by

$$I = I_0(\exp[eV/kT] - 1)$$

where V is the voltage across the junction. Show that when V is zero, the mean square value of the shot noise is

$$\langle i^2 \rangle = 4eI_0\,\Delta f$$

2

Optical Detectors

2.1 INTRODUCTION

Information about the physical world is presented to the observer in various forms − heat, light, sound and so on − and this information will also contain noise. In order to process this information, the carrier is usually first converted into an electrical signal by a transducer, after which amplification and various techniques to improve the ratio of signal to noise may be employed. The list of transducers is extremely long but probably the most important and widely employed is the optical transducer (hencefor- ward called a photodetector). In this chapter, the factors that characterize a photodetector will be discussed, namely its spectral response, its speed of response, and the noise added to the signal by the photodetector.

The electromagnetic spectrum is divided into several regions as shown in Fig. 2.1. The detectors discussed are those which detect ultraviolet, visible, and infrared radiation and are broadly called 'optical detectors'.

Two basic types of detector are in common use for sensing this optical radiation. These are:

1. thermal detectors, where the incident flux causes a temperature rise of the sensing element which can be measured in several ways. They normally operate at room temperature, have a response tailored to be independent of wavelength, and are relatively slow. Examples of these detectors are thermocouples, thermopiles, pyroelectric detectors and Golay cells.
2. photon detectors, where the individual photons comprising the optical flux interact directly with the electrons in the detector. This direct inter- action provides a very fast detector, making these detectors suitable for high frequency work, but the effect is also wavelength dependent. The photon detectors in common use are photomultipliers, photodiodes and photoconductors.

The difference in wavelength dependence between these two basic detector types is important. Photon detectors rely on the absorption of photon energy to promote a charge carrier across some energy barrier E_0. Photons of energy lower than this will effectively be ignored. This E_0 is characteristic of the type of detector and is usually fairly sharply defined. In terms of wavelength, photon detectors will show no response above the cut-off

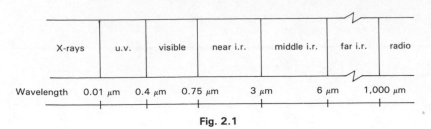

Fig. 2.1

wavelength

$$\lambda_0(\mu m) = \frac{1.24}{E_0(eV)}$$

For example, if $E_0 = 2$ eV, then the detector will not respond to light of wavelength greater than 0.62 μm (= 620 nm, in the red region of the visible spectrum). Thermal detectors, however, will not have this cut-off (see Fig. 2.2).

The spectral response shown in Fig. 2.2 may be given in either relative or absolute units. Relative response is normally given if interest is in comparing the output from the detector at different wavelengths. The units of absolute response depend on the type of detector employed; for example, for most photon detectors, the spectral response is given in amps of detector current per watt of incident optical energy.

The response time of a detector determines how rapidly the output of the photodetector will change when there is a step change in the input optical signal. Often, the bandwidth Δf of the detector is quoted rather than response time. Roughly speaking, this is the range of frequencies to which the deterctor will respond. Because of the connection, via the Fourier transform, of the time and frequency domain of the signal, there will be a direct relation between the response time and the bandwidth. Many photodetectors will produce an exponential rise to a step input, and, in this case, the response $R(\omega)$ will have the frequency dependence

$$R(\omega) = \frac{R(0)}{(1 + \omega^2\tau^2)^{1/2}}$$

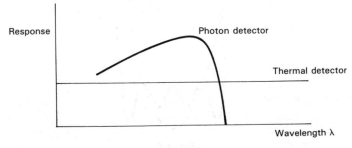

Fig. 2.2

The bandwidth is obtained when the response falls to $1/\sqrt{2}$ (≈ 0.707) of its value at zero frequency i.e.

$$\Delta f = \frac{1}{2\pi\tau}$$

If the response to the step is not exponential, the speed of response is still given by the time for the signal to fall to $1/e$ of its initial value and the bandwidth measured at the frequency at which the response falls to $1/\sqrt{2}$ of its zero frequency value.

Finally, the ability of the detector to measure a signal in the presence of noise must be considered. This will determine the smallest amount of optical which can be detected by the detector. This important topic must now be considered in some detail.

2.2 NOISE IN PHOTODETECTORS

In order to compare the performance of the various detectors, a 'figure of merit' must be introduced (Jones 1959, van Vliet 1958).

Probably the simplest is the 'minimum detectable power' W_m or noise equivalent power (NEP) which is the radiant flux in watts necessary to give a signal-to-noise ratio of 1 at the output of the detector. The detectivity D is the quantity most frequently used to describe the minimum detectable power and is the signal-to-noise per unit quantity of incident radiation:

$$D = \frac{(S/N)}{P} \ W^{-1}$$

As the noise depends on bandwidth Δf, and, for many detectors, on surface area A, it is customary to define the specific detectivity, D^* (called D star)

$$D^* = D(A\ \Delta f)^{+\frac{1}{2}} \ \text{cm} \ (\text{Hz})^{\frac{1}{2}} \ W^{-1}$$

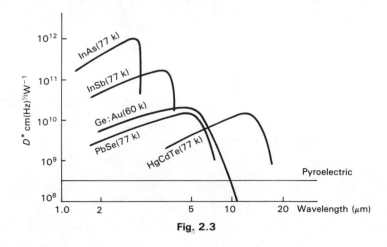

Fig. 2.3

When the spectral response and D^* are known, D^*_λ, the specific detectivity at a particular wavelength, may be evaluated. D^*_λ is a particularly useful number because its theoretical limit may be calculated.

D^* for some photodetectors and thermal detectors which are commercially available is given in Fig. 2.3.

2.2.1 Signal fluctuation and background fluctuation limits

The ultimate performance of detectors is achieved when there is no amplifier noise and no noise generated within the detector itself. For optical radiation, noise can arise from the radiating object (which we are interested in) and from the background. The radiation process is a quantum statistical process and so there will be fluctuations in the arrival of both signal and background photons; these random fluctuations will determine the minimum power detected by the detector.

Probably the most commonly encountered radiating background is that arising from thermal emission from surrounding bodies such as detector housings, and this usually at room temperatures. The amount of power radiated per unit area per unit solid angle per unit wavelength is given by Planck's equation:

$$W(\lambda, T) = \frac{2\pi h c^2}{\lambda^5} \frac{1}{[\exp(hc/\lambda kT) - 1]}$$

The shape of this curve is shown in Fig. 2.4.

The peak position of this curve, λ_{max}, is given by Wien's law

$$\lambda(\mu m) = \frac{2890}{T}$$

where T is the absolute temperature of the body. Thus, at room temperature, λ_{max} is at approximately 10 μm. Most of the energy (~ 75 percent) lies

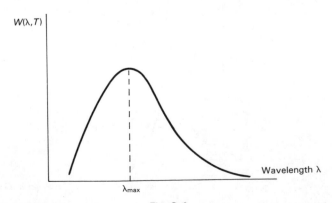

Fig. 2.4

at wavelengths greater than λ_{max}, that is, in the infrared and microwave regions. This has important consequences for the type of detector used. For visible and ultraviolet photon detectors, such as the photomultiplier, which have a sharp cut-off in their response, the background radiation will be unimportant. The minimum power detectable is now set by fluctuations in the arrival of signal photons and we are in the *signal fluctuation limit*. For infrared detectors, however, the background fluctuations will be the most important, and we now have the *background fluctuation limit*.

2.2.2 Photon noise in visible detectors

If we assume that the radiating body is black, then the fluctuations in the photon flux is given by Bose–Einstein statistics:

$$\langle \Delta n^2 \rangle = \langle n \rangle (1 + \langle n \rangle)$$

where

$$\langle n \rangle = \frac{1}{e^{hf/kT} - 1}$$

For optical photons at room temperature $hf \gg kT$, so that

$$\langle n \rangle \ll 1$$

It follows that, in this limit,

$$\langle \Delta n^2 \rangle = \langle n \rangle$$

For such a case, the photons are said to obey (classical) Poisson statistics and the photons are emitted completely independently of one another. The Poisson distribution function gives the probability of n photons arriving at a detector in the interval T as

$$p(n, T) = \frac{\langle n \rangle^n}{n!} e^{-\langle n \rangle}$$

where $\langle n \rangle$ is the mean number of photons arriving in the interval T. The mean square fluctuation in n is

$$\langle \Delta n^2 \rangle = \langle (n - \langle n \rangle)^2 \rangle$$
$$= \langle n^2 \rangle - \langle n \rangle^2$$

Observing that

$$\langle n^2 \rangle = \sum_{n=0}^{\infty} n^2 p(n) = \langle n \rangle + (\langle n \rangle)^2$$

then

$$\langle \Delta n^2 \rangle = \langle n \rangle$$

as stated above. The Poisson distribution is shown in Fig. 2.5.

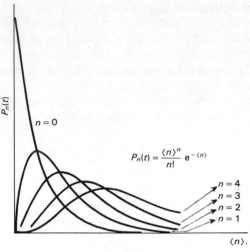

$$P_n(t) = \frac{\langle n \rangle^n}{n!} \, e^{-\langle n \rangle}$$

Fig. 2.5

When $n \gg 1$, Stirling's approximation for $n!$ can be used

$$n! = \sqrt{(2\pi n)} \, n^n \, e^{-n}$$

Writing $n = \langle n \rangle + \delta n$, it can be shown that

$$p(\delta n, T) = \frac{1}{\sqrt{2\pi\langle n \rangle}} \cdot \exp\left[\frac{-(\delta n)^2}{2\langle n \rangle}\right]$$

i.e. the fluctuations approach a Gaussian distribution about the mean with a standard deviation $\sigma = \sqrt{\langle n \rangle}$ (see Oliver 1965).

We can obtain an expression for the noise current obtained from a photodetector, due to photon noise, by a rough argument (Rose 1955). Imagine that we divide up the photon stream incident on the photodetector into equal time intervals T. The instantaneous photocurrent I_b will represent the superposition of a number of photons arriving at the photodetector over the small time intervals T. The mean photocurrent $\langle I_{ph} \rangle$ is obtained by averaging over a large number of these time intervals. Since the photon arrivals are random, independent events, we can write

$$\langle i^2 \rangle = \langle (I_b - \langle I_{ph} \rangle)^2 \rangle$$

$$= \frac{\alpha^2}{T^2}\langle (\Delta n)^2 \rangle$$

where α is the charge contribution per incident photon.

Now

$$\langle \Delta n^2 \rangle = \langle n \rangle$$

so that

$$\langle i^2 \rangle = \frac{\alpha^2}{T^2}\langle n \rangle = \frac{\alpha^2}{T} \frac{\langle I_{ph} \rangle}{e}$$

where e is the electronic charge, and

$$T = \frac{1}{2\Delta f}$$

where Δf is the bandwidth of the detector.

Hence,

$$\langle i^2 \rangle = \frac{2\alpha^2}{e} \langle I_{ph} \rangle \, \Delta f$$

which is of shot noise form. In particular, for a quantum efficiency of 1,

$$\alpha^2 = e^2$$

and

$$\langle i^2 \rangle = 2e \langle I_{ph} \rangle \, \Delta f$$

2.2.3 Minimum power detectable with photodetectors

The random emission of photons will set the basic limit to the minimum power detectable by a photodetector. Suppose that photons are arriving at a photodetector and further suppose that the photodetector and any following electronics add no noise to the signal. The observation of intensity is achieved by dividing the observation time up into elements of length T and counting the number of pulses produced from the photodetector. If there are N photons per second arriving at the photodetector of quantum efficiency η, then the number of photoelectrons generated is

$$S = N\eta T$$

If the photon distribution obeys Poisson statistics, the noise is

$$\sqrt{\langle \Delta s^2 \rangle} = \sqrt{s} = \sqrt{N\eta T}$$

The minimum number of photons which can be detected occurs when the signal equals the noise:

$$N\eta T = \sqrt{N\eta T}$$

or

$$(N)_{min} = \frac{1}{\eta T}$$

In terms of optical power, the minimum detectable power due to the statistical nature of photon emission is

$$(P)_{min} = \frac{hf}{\eta T} \approx \frac{2hf \, \Delta f}{\eta}, \quad \Delta f = \text{bandwidth}$$

For the infrared region of the spectrum where $hf \approx kT$ (i.e. at wavelengths of around 50 μm at room temperature), no simplifications can be made and the full expression for the fluctuations in the number of photons must be used. However, it is possible to obtain an expression for the minimum power detectable for thermal detectors. The temperature of the thermal detector will be determined by a statistical interchange of energy between it and its surroundings. Its temperature will thus fluctuate in a random fashion about a mean value. The minimum power detectable (Putley 1973) is then given by

$$P_{min} = 16Ak\sigma\varepsilon T^5 \, \Delta f$$

where A is the detector area, σ is Stefan's constant, and ε the emissivity at a temperature T. For a detector with $A = 0.01$ cm^2, $T = 300$ K, $\Delta f = 1$ Hz:

$$P_{min} \approx 5 \times 10^{-12} \text{ W}$$

For photon detectors of comparable bandwidth:

$$P_{min} = 10^{-19} \text{ W}$$

while in the microwave region where $hf \ll kT$:

$$P_{min} \approx 4 \times 10^{-21} \text{ W}$$

The high value for thermal detectors is because they are sensitive to the whole of the background radiation falling on them. Photon detectors have cut-off frequencies and microwave detectors have narrow pass-bands.

2.3 PHOTON DETECTORS

Photon detectors are generally divided into two broad areas depending on whether the charge carriers are emitted externally from a solid (external photoelectric device) or whether they are freed but remain inside the solid (internal photoelectric device). In the former category, the best known device is undoubtedly the photomultiplier, while in the latter, several devices are commonly available, such as photodiodes and photoconductors.

2.4 EXTERNAL PHOTOELECTRIC DEVICES

The photomultiplier is an example of a photodetector which uses the external photoemissive effect where photons incident upon a metallic surface will liberate electrons. There is a certain minimum amount of energy needed to liberate an electron from the surface of a metal which is known as the work function. If a photon of energy greater than this is incident on the metal, an electron may be emitted from the surface; this is known as photoemission. The work function of metals typically lies in the range 1.5–3.0 eV so that the cut-off wavelength λ_0 lies in the range 400–800 nm, i.e. in the visible region of the optical spectrum.

There are several requirements for materials if they are to be useful as photoemissive surfaces. Firstly, they should have relatively low work functions. Secondly, photoelectrons excited within the bulk of the photoemitter should be able to reach the surface with a minimum of energy loss in order to obtain a high efficiency device. Thirdly, there should be enough electrical conductivity to permit neutralization of any charge accumulated on the surface due to the photoemission.

Commercially available photocathodes are themselves divided into two broad areas. So-called classical photoemitters are the most readily available and are based on alkali metals or monomolecular layers of alkali metal adsorbed on the surface of such metals as silver or tungsten. The earlier photocathodes were assigned S numbers dependent on their spectral response. For example, the S1 photocathode consisting of silver with caesium oxide (Ag/CsO) has an active range extending up to 1,000 nm while the S20 multialkali photocathode combines high sensitivity with operation extending to the near infrared. Some bialkali photocathodes were introduced after the S number system and are simply referred to by their composition, RbCs bialkali photocathodes have a broader spectral response

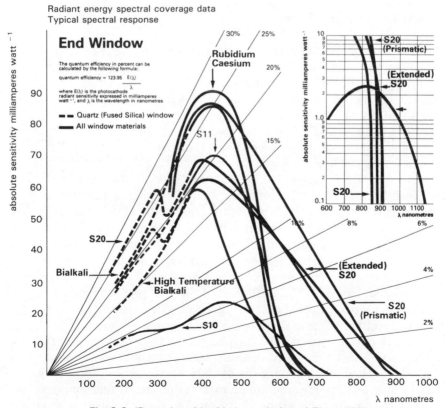

Fig. 2.6 (Reproduced by kind permission of Thorn EMI)

than S20 photocathodes but if a low dark count is required (see later for more information on dark count) KCs cathodes are preferred. KNaSb is a high temperature bialkali, capable of operating at temperatures up to $175°C$. The second class of photocathodes are known as 'negative electron affinity' or NEA devices. These combine high sensitivity in the visible region of the spectrum with good response in the infrared by coating a semiconductor such as gallium arsenide (GaAs) with a film of caesium oxide (Cs_2O). Such techniques will produce photocathodes sensitive in the infrared out to 1.2 μm; beyond this wavelength, no stable films are available. A comparison between several different photocathode materials is given in Fig. 2.6.

For detection in the ultraviolet with significant efficiency, several photocathode materials are available. Materials such as potassium bromide KBr and caesium iodide CsI have very limited responses (between $\sim$ 120 nm and $\sim$200 nm) and are particularly useful for the detection of ultraviolet light in the presence of visible photons. Other photocathode materials such as caesium telluride and rubidium telluride have wider responses (100–350 nm).

2.4.1 The photomultiplier

The output from a simple photoemissive device will generally be small and some form of amplification will be needed. The most popular means of achieving this is the photomultiplier, which uses an internal gain mechanism. In the photomultiplier, the photocathode is contained in an evacuated sealed tube together with a number of electron-multiplying electrodes or dynodes and a final, electron-collecting, anode. The tube material is usually a boro-silicate glass which will transmit light with wavelengths as low as about 300 nm; to extend this further into the ultraviolet, materials such as sapphire or magnesium fluoride can be used. The photocathode is specified by several parameters. The spectral sensitivity has already been discussed. The quantum efficiency of the photocathode gives the percentage number of photoelectrons emitted from the photocathode per incident photon. The radiant sensitivity specified at a particular wavelength is the photocurrent in milliamps per watt of incident optical power. Both these quantities are specified at a particular wavelength and it is customary to give the cathode sensitivity in μA/lumen. It is the photocurrent from the photocathode per incident light flux from a tungsten filament lamp operated at a temperature of 2856 K.

Photoelectrons, ejected from the photocathode, are accelerated towards the dynodes, each one at a successively higher potential with respect to the cathode. Each electron striking the dynode releases several secondary electrons which, in turn, are accelerated towards the next dynode and so on until the electrons are collected by the anode. The anode behaves as an almost

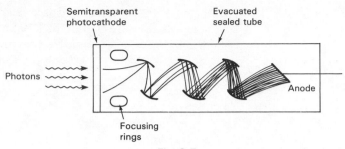

Fig. 2.7

pure current source. If there are N dynodes, each with an average gain per dynode of δ (the secondary emission process is a statistical process), the overall gain of the photomultiplier tube will be

$$G = \delta^N$$

which is about 10^6 for $N = 10$ and $\delta = 4$.

A schematic diagram of a photomultiplier is shown in Fig. 2.7.

If a pulse of light is incident on the photocathode, the electrons arising at the anode show an approximately Gaussian spread in arrival times (see Fig. 2.8).

The spread is chiefly caused by fluctuations in the times of flight of photoelectrons and secondary electrons due to their different initial velocities and trajectories. The anode pulse rise time is the time for the output pulse to rise from 10 to 90 percent of the peak output. The electron transit time is the time between the photon pulse at the photocathode and the maximum of the anode current pulse. The spread of the electrons arriving at the anode determines the speed of response of the photomultiplier (and hence its bandwidth) and also the time resolution capabilities of the device.

The dynode potentials are usually provided by a circuit of the type shown in Fig. 2.9).

Here, the normal configuration of the grounded anode is employed. Photomultipliers can be operated with the cathode at ground and the anode at a positive high voltage, which can result in a reduction of noise (see the

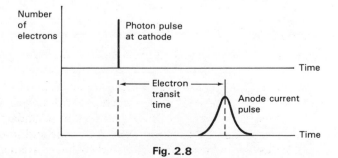

Fig. 2.8

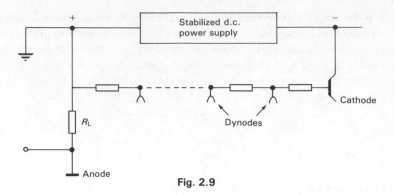

Fig. 2.9

section on dark current later in the chapter). Since the anode acts as an almost pure current source, a resistor R_L can be employed as a load to convert this photocurrent to a voltage. There are practical limits to the size of the load and very large values can have two undesirable effects. Firstly, if the voltage drop across the load becomes too large, the potential between the last dynode and the anode can fall and can affect the linear relation between anode current and incident photon flux. This can be avoided by restricting the value of the load so that only a few volts are dropped across it. Secondly, the load resistor will affect the frequency response of the photomultiplier. This is due to the unavoidable stray capacitances which exist across the load (Fig. 2.10).

The $R_L C_S$ combination will have a bandwidth given by

$$\Delta f = \frac{1}{2\pi C_S R_L}$$

Thus, suppose we have a fast photomultiplier with a bandwidth of, say, 100 MHz, with a load resistance of 10 MΩ and stray capacitance of 1 pF. The bandwidth of the $R_L C_S$ combination is about 15 kHz, which has severely limited the bandwidth of the total system. If speed of response is important, the load resistance should be kept as small as possible and all stray capacitances minimized.

The power supply itself has to be well stabilized and this can be demonstrated by a simple argument. The average gain per dynode can be written

$$\delta = \text{const.}\ E^\alpha$$

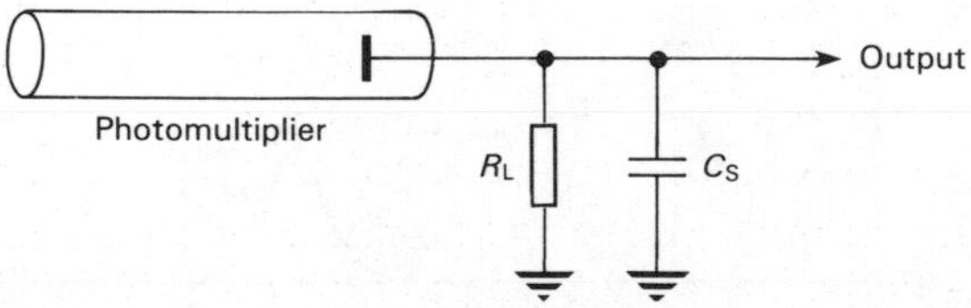

Fig. 2.10

where E is the interstage voltage and α is a coefficient whose value is determined by the dynode material (α usually lies between 0.7 and 0.8). This gain per stage is almost linear up to 100 V but the nonlinearity becomes marked above this. Since the overall gain of the photomultiplier is

$$G = \delta^N$$

where N is the number of dynode stages,

$$G = (\text{const. } E^\alpha)^N$$
$$= \text{const. } E^{\alpha N}$$

Now, the voltage per stage E is proportional to the overall voltage between anode and cathode V, hence

$$G = KV^{\alpha N}$$

Differentiating

$$dG = K\alpha NV^{(\alpha N - 1)} \, dV$$

or

$$\frac{dG}{G} = \alpha \, \frac{N \, dV}{V}$$

Thus, fluctuations in the gain are a factor αN times the fluctuation in the photomultiplier voltage. Since αN is around 10 for most photomultipliers, this implies that the power supply should be about ten times better than the desired output stability.

A significant part of the time spread of electrons arriving at the anode after a very short pulse of light is due to electrons arriving at the first dynode at different times and, when the time resolution of photomultiplier is important, a zener diode is used between the cathode and first dynode to maintain the potential between them despite variations in dynode chain current.

When the photomultiplier is used to detect pulsed light, the final two or three dynodes are decoupled by capacitors in order to maintain dynode potentials at a constant value while delivering high peak currents.

The traditional dynode materials are caesium antimony (CsSb) or beryllium copper (BeCu). Newer materials such as caesium activated gallium phosphide (GaP) result in better time resolution than the traditional dynode materials. There are several popular physical arrangements of the dynodes: the venetian blind, box linear focused, and circular cage focused types (Fig. 2.11). The first two types do not use any electron focusing and hence the spread in transit times of the electrons from cathode to anode is significantly larger in these arrangements than in the focused types. Thus the focused dynode arrangements are more suitable for high frequency operation. For example, a typical photomultiplier with venetian blind

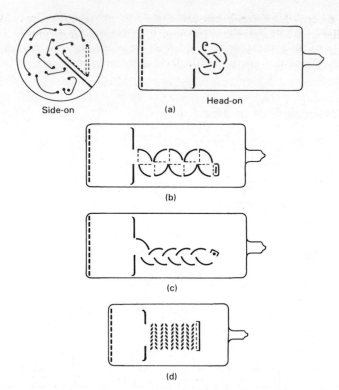

Fig. 2.11 Types of electron multiplier: (a) circular cage type; (b) box-and-grid type; (c) linear focused type; (d) Venetian blind type

dynode structure (EMI 9956) will have a spread in transit time of 25 ns while one with a linear focused dynode structure (EMI 9954) has a spread in transit times of 3 ns.

2.4.2 Noise and the minimum detectable signal in photomultipliers

In this section we shall consider the causes of noise in photomultipliers and so develop a noise model for them. This will allow us to determine the minimum detectable signal in photomultipliers.

The major source of noise in photomultipliers is thermionic emission (also known as dark current) from the photocathode (Sharp 1964). This is a statistical process, and the shot noise associated with this dark current as measured at the anode is

$$\langle i^2 \rangle = G^2 e \langle I_d \rangle \, \Delta f$$

where $\langle I_d \rangle$ is the mean value of the dark current due to the thermionic emission. It is this random variation of dark current which is the limitation to the detection of weak optical signals in photomultipliers. We have seen earlier in this chapter that the noise due to signal fluctuations is also of shot noise form; it is impossible to determine whether the resultant shot noise in a photomultiplier is due to photon noise or thermionic emission.

There will also be thermionic emission from the dynodes which will contribute to the shot noise at the anode. The mean anode current due to the jth dynode is $\delta^j \langle I_d \rangle$ and the shot noise at the anode due to this dynode is

$$\langle i_j^2 \rangle = 2e\delta^j \langle I_d \rangle \, \Delta f (\delta^{N-j})^2$$

where N is the total number of dynode stages. The total shot noise due to the dynodes is then

$$\sum_{j=1}^{N} \langle i_j^2 \rangle = 2e\langle I_d \rangle \, \Delta f(\delta(\delta^{N-1})^2 + \delta^2(\delta^{N-2})^2 + \ldots + \delta^N)$$

$$= 2e\langle I_d \rangle \, \Delta f(\delta^N)^2 \left(\frac{1}{\delta} + \frac{1}{\delta^2} + \ldots + \frac{1}{\delta^N} \right)$$

$$= 2e\langle I_d \rangle \, \Delta f \cdot G^2 \left(\frac{1}{\delta} + \frac{1}{\delta^2} + \ldots + \frac{1^N}{d^N} \right)$$

(Since the gain G of the photomultiplier is $G = \delta^N$.)

The total dark noise at the anode can be written

$$\langle i^2 \rangle = 2e\langle I_d \rangle G^2 \, \Delta f \left(1 + \frac{1}{\delta} + \frac{1}{\delta^2} + \ldots + \frac{1}{\delta^N} \right)$$

This expression is sometimes written

$$\langle i^2 \rangle = 2e\langle I_d \rangle \, \Delta f G^2 F(G)$$

where $F(G)$ is known as the excess noise factor, defined as the ratio of the signal-to-noise ratio at the input of the multiplication process to that at the output. For photomultipliers with $\delta \approx 10$

$$F(G) \rightarrow 1$$

In the presence of light with an average optical power P, there will be an average photocathode current

$$I_s = \frac{e\eta}{hf_s} (P_{ph} + P_B)$$

where f_s is the frequency of the optical signal of power P_{ph} and P_B is the background optical power. The fluctuations of the optical flux produce a shot noise contribution at the anode of

$$\langle i_s^2 \rangle = 2e\langle I_{ph} \rangle G^2 F \, \Delta f$$

Finally, the anode resistor R will produce Johnson noise with mean square current

$$\langle i_R^2 \rangle = \frac{4kT}{R} \, \Delta f$$

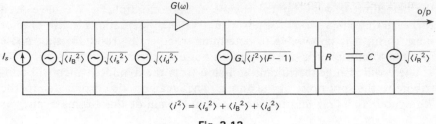

$$\langle i^2 \rangle = \langle i_s^2 \rangle + \langle i_B^2 \rangle + \langle i_d^2 \rangle$$

Fig. 2.12

The model for the photomultiplier is shown in Fig. 2.12, and this is a general model for photodetectors which will be used throughout this chapter. Here $\langle i_d^2 \rangle$ is the mean square dark current from the photocathode and C represents any stray capacitance across R (or any input capacitance from a following amplifier). $G(\omega)$ is the (frequency-dependent) gain of the photomultiplier, and F the excess noise factor.

The cathode current will produce an anode current

$$I_{sa} = I_s G(\omega)$$

$G(\omega)$ can be written in terms of the anode pulse rise time, defined earlier as

$$G(\omega) = \frac{G(0)}{(1 + \omega^2 \tau_r^2)^{\frac{1}{2}}}$$

where $G(0)$ is the midband gain obtained at frequencies such that $\omega \tau_r \ll 1$. The anode current will develop a voltage across R given by

$$V_{out} = I_s G(\omega) \frac{R}{(1 + \omega^2 R^2 C^2)^{\frac{1}{2}}}$$

$$= \frac{e\eta P_s}{hf_s} \cdot \frac{R}{(1 + \omega^2 R^2 C^2)^{\frac{1}{2}}} \cdot \frac{G(0)}{(1 + \omega^2 \tau_r^2)^{\frac{1}{2}}}$$

The signal-to-noise *power* at the output of the photomultiplier is

$$\frac{S}{N} = \frac{(e\eta P_{ph}/hf_s)^2 G^2(\omega)}{\left\{ 2G^2(\omega)e\left[\langle I_d \rangle + \frac{\eta e}{hf_s}(P_{ph} + P_B) \right] \Delta f F + \frac{4RT \, \Delta f}{R} \right\}}$$

This can be solved to obtain the amount of signal power necessary to achieve a signal-to-noise power ratio of unity:

$$(P_s)_{min} = \frac{\Delta f \, hf_s F}{\eta} \left\{ 1 + \left[1 + \frac{1}{\Delta f F^2} \left\{ \left(\frac{\eta P_B}{hf_s} + \frac{2\langle I_d \rangle}{e} \right) F + \frac{4kT}{e^2 G^2 R} \right\} \right]^{\frac{1}{2}} \right\}$$

The square root term in the bracket represents the combined effect of all noise sources (dark current, load resistor) exlucing the shot noise of the signal. For the photomultiplier $G \approx 10^6$ and $F \approx 1$, and as discussed earlier, for visible photodetectors, we can ignore the background flux. Thus, the minimum detectable power is

$$(P_{ph})_{min} = \frac{\Delta fhf_s}{\eta} \left\{ 1 + \left[1 + \frac{2\langle I_d \rangle}{e \, \Delta f} \right]^{\frac{1}{2}} \right\}$$

Typical values for a photomultiplier are

$$\eta = 10\%$$
$$\langle I_d \rangle = 10^{-15} \, \text{A}$$

and assuming photons of wavelength $\lambda = 500$ nm

$$f_s = 6 \times 10^{14} \, \text{Hz}$$

and we will choose $\Delta f = 1$ Hz.

In this case

$$(P_{ph})_{min} \approx \frac{\Delta fhf_s}{\eta} \left(\frac{2\langle I_d \rangle}{e \, \Delta f} \right)^{\frac{1}{2}}$$

and it is the dark current and its fluctuations which are the limiting factors in the sensitivity of the photomultiplier. If the dark current could be eliminated, it would be the fluctuations in the signal that would be the limiting factor. This is known as the quantum limit of detection and

$$(P_{ph})_{min} = 2 \, hf_s \frac{\Delta f}{\eta}$$

For the 'typical' values

$$(P_{ph})_{min} = 4 \times 10^{-18} \, \text{W}$$

Since P_s/hf_s is the number of photons arriving at the detector per second, the quantity

$$\frac{(P_{ph})_{min}}{hf_s \, \Delta f} = \left(\frac{2}{\eta} \right)$$

is the number of photons per second per unit bandwidth needed to achieve a power signal-to-noise ratio of unity.

2.5 INTERNAL PHOTOELECTRIC DEVICES

An ideal semiconductor can be described by an energy level diagram which has a lower band of electronic states, called the valence band, which states are completely filled at absolute zero, separated by an energy gap from a higher band of allowed states which are completely empty at absolute zero. In the ideal case, there are no allowed states in the energy gap. At absolute zero the ideal semiconductor is a perfect insulator, but, at finite temperatures, electrons can be thermally excited from the valence band to the conduction band, leaving 'free' holes in the valence band. Both electrons and holes can then contribute to electrical conduction. This is called intrinsic excitation of carriers. Intrinsic excitation can also be produced by absorption of photons, thus producing photoexcited carriers which can contribute to the electrical conductivity of the semiconductor. Such an internal photoelectric effect is the basis of such photodetectors as the photoconductor and the photodiode. The minimum wavelength to produce intrinsic photoconductivity is

$$\lambda_0(\mu m) = \frac{1.24}{E_g(eV)}$$

where E_g is the minimum energy gap between the conduction and valence band in electron volts. A schematic diagram illustrating intrinsic absorption is shown in Fig. 2.13.

The value of the band gap and threshold wavelengths of some semiconductors is given in Table 2.1.

The photo response of the semiconductor can be extended to longer wavelengths by doping it with impurity atoms, giving rise to extrinsic conductivity. This can be best understood by an example. Consider the addition of a small concentration of phosphorus atoms to a crystal of silicon in the proportion of 1 phosphorus atom to 10^5 silicon atoms. Silicon is a tetravalent covalently bonded solid. If a phosphorus atom, which is pentavalent, replaces a silicon atom, there will be one valence electron left over after the impurity atom has been accommodated into the structure. The binding

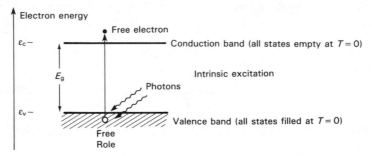

Fig. 2.13

Table 2.1 Band gaps and threshold wavelengths for some semiconductors

Semiconductor	Band gap(eV)	$\lambda_0(\mu m)$
Si	1.14	1.1
Ge	0.67	1.85
InSb	0.23	5.4
InAs	0.33	3.76
GaAs	1.4	0.88
CdS	2.42	0.51
Pbs	0.34	3.7

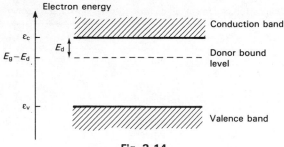

Fig. 2.14

energy of this extra electron is much less than the band gap of the pure silicon (0.045 eV). At any temperature, a fraction of these impurities will be ionized so that there will be net excess of electrons in the semiconductor (remembering that there will also be free electrons and holes due to intrinsic excitation) and the semiconductor is known as n-type extrinsic. In this case, the impurity atoms are known as donors. The energy band diagram is shown in Fig. 2.14.

The bound donor electrons may also be freed by absorption of a photon. In this case, the cut-off wavelength for photoconductivity is

$$\lambda_0 = \frac{hc}{E_d}$$

In a similar fashion, we can add trivalent impurity atoms to the lattice. In this case, a hole will be bound to the impurity. An electron can be excited from the valence band into the hole, leaving a free hole in the valence band. We now have a p-type extrinsic semiconductor, and the impurity is called an acceptor. The energy band diagram is shown in Fig. 2.15. The cut-off wavelength for photoconductivity is

$$\lambda_0 = \frac{hc}{E_a}$$

The ionization energy of some impurities in silicon and germanium are given in Table 2.2 (Schultz and Morton 1955).

These ionization energies are very small compared with the thermal

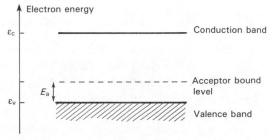

Fig. 2.15

Table 2.2 Ionization energy of impurities in silicon and germanium

Host	Impurity	Ionization energy (eV)	$\lambda_0 (\mu m)$
Ge	B(a)	0.0104	119.0
	Al(a)	0.0102	121.0
	Ga(a)	0.0108	115.0
	P(d)	0.0120	103.0
	As(d)	0.0127	98.0
	Sb(d)	0.0097	127.0
Si	B(a)	0.0450	27.5
	Al(a)	0.0570	21.7
	Ga(a)	0.0650	19.0
	In(a)	0.1600	8.0
	P(d)	0.0390	31.0
	As(d)	0.0490	25.0
	Sb(d)	0.0390	31.0

energy at room temperature ($kT \approx 0.025$ eV at $T = 300$ K). At room temperature, a significant fraction of these impurities will be ionized, leading to reduced efficiency for photoconductivity and a more noisy device. For these reasons, such extrinsic detectors have to be cooled, and this cooling requirement is more stringent than with thermal detectors, which can sometimes be a disadvantage.

The doping of semiconductors is not the only available method for tailoring the optical absorption of semiconductors. Certain compound semiconductors will exist in stable solid solution with others; for example, mercury telluride HgTe and cadmium telluride CdTe. The band gap of the compound cadmium mercury telluride ($Cd_xHg_{1-x}Te$) depends on the relative proportions of HgTe and CdTe (i.e. Eg depends on x) and the material is a useful photodetector in the 5–14 μm region of the infrared. Similarly, $Pb_xSn_{1-x}Te$ is used in this region. The importance of optical communications has led to the development of other compound materials. For example, gallium aluminum arsenide $Ga_xAl_{1-x}As$ is a well-developed photodetector having a useful operating range between 650 and 900 nm, while the push to optical communications beyond 1 μm is seeing the introduction of materials like $In_xGa_{1-x}As$ and $GaAs_{1-x}Sb_x$.

2.6 OPTICAL ABSORPTION IN SEMICONDUCTORS

We have now seen that there are two major ways of obtaining photogenerated carriers in a semiconductor: excitation across the band gap, and excitation from impurities. It is important in the design and choice of a photodetector to know roughly the magnitude of the optical absorption coefficient for the different processes.

Consider a slab of semiconductor of thickness d with light of intensity incident upon it. If the semiconductor has a reflection coefficient R_0, then, at some distance x within the slab, the light intensity will be

$$I(x) = (1 - R_0)I_0 \, e^{-\alpha x}$$

The ratio of the number of photons absorbed in the slab to the initial number of photons is called the quantum efficiency η and is given by

$$\eta = (1 - R_0)[1 - \exp(-\alpha d)]$$

α is known as the absorption coefficient and its value will now be discussed.

2.6.1 Intrinsic semiconductors

Absorption across the band gap in a semiconductor is somewhat more complex than we have so far discussed. This is because optical transitions may involve photons only (direct) or may be moderated by lattice vibrations (indirect). This is best discussed with reference to Fig. 2.16, which shows the energy band structure as a function of wavevector for some hypothetical semiconductor (Kittel 1968).

In the direct transition A, the photon is absorbed by the electron such that the electron suffers negligible change in wavevector, since the photon itself brings negligible momentum. For the indirect transition B, conservation of momentum in the crystal means that momentum must be supplied by lattice vibrations. Effectively, B is a two-step process and we may expect the absorption coefficient in A to be greater than in B. Typically, α for a direct transition will have a value of $\leqslant 10^6 \, \text{cm}^{-1}$ so that photons will be strongly absorbed in the surface layers of the semiconductor. For the indirect transition, α has values of $\leqslant 10^4 \, \text{cm}^{-1}$.

So far in this discussion we have only referred to a hypothetical semiconductor. In practice, we can distinguish two types of semiconductor. The first has the same value of wavevector for the lowest energy state in the conduction band as for the highest energy state in the valence band; for example, indium antimonide InSb. In the second type, this is not so; for example, germanium and silicon. We should notice in the second type that, if we make the photon frequency large enough, we will eventually get a direct transition, so that the absorption coefficient will rise from its

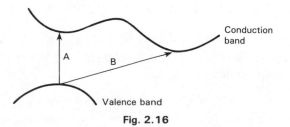

Fig. 2.16

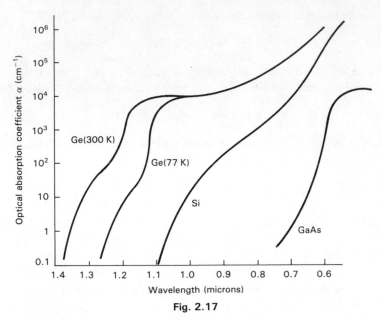

Fig. 2.17

10^4 cm^{-1} value at lower photon frequencies to 10^6 cm^{-1} higher photon frequencies (Fig. 2.17).

2.6.2 Extrinsic semiconductor

To first order, the absorption coefficient from impurities will depend on their density. This is usually very small compared with the density of intrinsic atoms and we can thus expect the absorption coefficient to be very small, with values of 10 cm^{-1} and less, e.g. in Ge containing $\sim 10^{16}$ atoms cm^{-3}, $\alpha \approx 0.15$ cm^{-1}.

The above discussion is by no means exhaustive; we have not discussed 'forbidden' transitions, excitons or surface states. The reader is directed to texts on semiconductors if it is desired to pursue this section further (Smith 1959, Kittel 1968).

2.7 PHOTOCONDUCTIVITY AND PHOTOCONDUCTIVE DETECTORS

Consider a slab of semiconductor that is irradiated by light which creates Δn electrons per unit volume and Δp holes per unit volume. The conductivity of the semiconductor is then

$$\sigma = e(n_0 + \Delta n)\mu_e + e(p_0 + \Delta p)\mu_h$$

where n_0, p_0 are the thermally generated carrier densities and μ_e, μ_h are

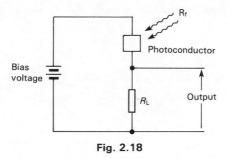

Fig. 2.18

electron and hole mobilities respectively. If the semiconductor is connected to an external circuit via contacts which essentially provide no barrier to current flow (ohmic contacts) then an externally applied electric field can extract the photogenerated carriers. We thus have a photon detector. A typical experimental arrangement is shown in Fig. 2.18.

There are two common methods used for biasing photoconductors:

1. The load resistor R_L is placed in series with the photodetector and has the same resistance. This maximizes the power delivered to the load. In fact, we more frequently want to maximize the *change* in load power due to a change in detector resistance. In this case, it can be shown that the load resistance R_L should be half the detector resistance.
2. A constant d.c. bias current is used with a lower impedance, current-sensitive preamplifier. In this case, the signal voltage obtained is directly proportional to the change in photoconductor resistance.

2.7.1 Recombination of photocarriers

Our discussion of the process of photoconductivity is not yet complete since we have not considered the important problem of what exactly happens to the photogenerated carriers after they have been freed. Let us consider the case of an intrinsic semiconductor and, to simplify the mathematics, assume that only one carrier type (say holes) makes a significant contribution to the current. These holes are constantly being generated by thermal and optical means and in our (hypothetically pure) semiconductor are removed from the conduction process when they combine with electrons. We may write, for the rate of change of holes per unit volume

$$\left(\frac{\mathrm{d}p}{\mathrm{d}t}\right) = R_{th} + R_{ph} - \left(\frac{p}{\tau}\right)$$

where τ is the carrier lifetime, R_{th} is the thermal generation rate, and R_{ph} is the photogeneration rate.

If we assume for simplicity a uniform generation of carriers throughout the volume of the semiconductor, then

$$R_{\mathrm{ph}} = \frac{\alpha I}{h\nu}$$

In the absence of light, and in equilibrium

$$R_{\mathrm{th}} = \frac{p_0}{\tau}$$

where we make the further simplifying (and not totally justified) assumption that τ is independent of carrier concentration.

Thus

$$\frac{\mathrm{d}p}{\mathrm{d}t} = \frac{p_0 - p}{\tau} + \frac{\alpha I}{h\nu}$$

At equilibrium in the presence of light, $\dfrac{\mathrm{d}p}{\mathrm{d}t} = 0$ and

$$\frac{\Delta p}{\tau} = \frac{p - p_0}{\tau} = \frac{\alpha I}{h\nu}$$

Suppose we switch on the light at $t = 0$. p may be written

$$p = p_0 + \frac{\alpha I \tau}{h\nu} (1 - e^{-t/\tau})$$

as may be verified by direct substitution. The quantity $\dfrac{\alpha I \tau}{h\nu}$ is simply the excess equilibrium hole density Δp and so

$$p = p_0 + \Delta p (1 - e^{-t/\tau})$$

Similarly, if we switch off the light at some time t_0 later, the hole density will decay according to

$$p = p_1 - \Delta p\, e^{-(t-t_0)/\tau}$$

where p_1 is the density of holes at t_0. Thus, the lifetime τ acts as a time constant or response time for the photoconductor.

In semiconductors, however, direct electron–hole recombination is not the only means of relaxing the electron–hole free carrier density. Impurities may also play a major role here. We can distinguish two main effects. In the first case, we can have recombining traps. One of the carriers will be captured by the trap which then subsequently captures the opposite type of carrier, bringing about recombination. In this case, the analysis is exactly the same as above but the time constant is no longer equal to τ. The second type of trap is called a nonrecombining trap. When this trap captures a carrier type, the cross-section for capture of the opposite carrier type is extremely small, effectively increasing the time it is free. This means that we have increased the photosignal for a given photon flux. The deliberate addition of traps to achieve this effect is known as sensitizing.

This increase in gain is, however, achieved at the expense of the response time of the detector. The above discussion has implied that the response time of the photoconductor is identical with the recombination time. In practice, this is not found to be so due to the inevitable existence of unintentional impurities which have very low binding energies. When the illumination is removed from the photoconductor, these traps slowly release any charge carriers captured by them. It can be shown that the response time τ_0 is related to the recombination time τ by

$$\tau_0 = \tau \left(1 + \frac{n_T}{n} \right)$$

where n_T = density of shallow traps per unit energy internal, and n = density of free carriers.

When $n \gg n_T$, the photocurrent will fall out with time constant τ, but eventually the response time will be

$$\tau = \tau_0 \frac{n_T}{n}$$

This demonstrates the importance of purity and perfection of the material needed to obtain fast response times.

2.7.2 Gain in photoconductors

An expression for the midband gain of a photoconductor can also be obtained by simple arguments. If the photoconductor has an area A and the distance between the electrodes is l, the photocurrent flowing is

$$I = \frac{VA}{l} \Delta \sigma$$

where $\Delta \sigma$ is the change in conductivity of the semiconductor due to the light and V is the voltage across the photoconductor. Using the expression for $\Delta \sigma$

$$I = \frac{VA}{l} \cdot e\mu_h \, \Delta p$$

$$= \frac{VA}{l} \cdot e\mu_h \frac{\alpha I \tau}{hf}$$

The ratio $\alpha I / hf$ represents the number of holes N_h per unit volume produced by the incident light per second. Therefore

$$I = \frac{VA}{l} \, e\mu_h \, \frac{N_h \tau}{Al}$$

Since μ_h is the velocity of the holes per unit electric field

$$\mu_h = \frac{v}{E} = \frac{v}{V/l}$$

we can further write

$$I = \frac{ve}{l} N_h \tau$$

Finally, if we let the number of carriers collected at the electrodes per second be N_I, then

$$I = N_I e = \frac{v}{l} eN_h\tau$$

Thus, the ratio of the number of holes collected at the electrodes per second to the number of holes generated per second by the light represents a gain G given by

$$G = \frac{N_I}{N_h} = \frac{\tau}{l/v} = \frac{\tau}{T}$$

where T is the transit time of carriers between the electrodes.

A photoconductor can have a gain of greater than unity if $\tau > T$. This means that the carrier lifetime is longer than the drift time of carriers between the electrodes. When a carrier reaches the contact and passes into the external circuit, charge neutrality ensures that another carrier must enter the photoconductor from the other electrodes. This is the physical interpretation of the gain of a photoconductor.

For a given incident light level, it would seen that the gain can be increased by decreasing T (either by increasing the voltage or making the crystal thinner). However, the eventual build-up of space charge will occur, because the contacts cannot remove or supply charge carriers freely, which will reduce the photocurrent. The minimum value of T is given by

$$T_{min} = \frac{lC}{\mu NeA}$$

where C is the capacitance of the device. Recognizing that $l/\mu NeA$ is simply the resistance of the photoconductor R_0, then

$$T_{min} = R_0 C$$

and so is the RC constant of the photodetector.

For a photoconductor operating within a bandwidth Δf, to obtain a flat frequency response and minimum phase distortion, τ should be as small as possible. This will, however, reduce the current gain. The value of τ is first decided on bandwidth requirements and then the gain is achieved by minimizing the transit time T by using high bias voltages and small dimensions in the direction of current flow. There is still the problem that the

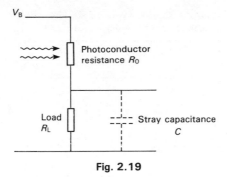

Fig. 2.19

photodetector feeds a load resistor R_L which will have some stray capacitance associated with it, in addition to the capacitance of the photodetector itself (Fig. 2.19).

If this capacitance is lumped together to give some value C, the bandwidth due to the RC effect is

$$\Delta f = \frac{1}{2\pi R_{||}C}$$

where

$$\frac{1}{R_{||}} = \frac{1}{R_L} + \frac{1}{R_0}$$

For wideband signals, this requires small values of R_L and, as will be seen later, leads to reduced sensitivity of the photoconductor since the thermal noise of the load becomes very large. High sensitivity and high bandwidth can only be achieved in photoconductors by the use of heterodyne detection (see Chapter 5).

2.7.3 Materials for photoconductive detectors

Photoconductive detectors can be made both intrinsic and extrinsic. Silicon and germanium intrinsic photoconductive detectors (Schultz and Morton 1955) have a cut-off at 1.1 μm and 1.7 μm respectively, and the response of these detectors has been extended to longer wavelengths by doping. The range of germanium extrinsic detectors is exhaustive, including the dopants mercury (Ge:Hg with a response out to 12 μm), gold (Ge:Au with a response out to 10 μm), and copper (Ge:Cu with a response out to 30 μm). Silicon extrinsic detectors have also been produced with impurities such as indium In and gallium Ga (see Table 2.3). The interest in extrinsic silicon has arisen because of the desire to construct an integrated sensor (detector, load, and preamplifier) onto a single silicon substrate. Indium antimonide has been used as a photoconductive detector at wavelengths of 3 to 5 μm

Table 2.3 Some extrinsic silicon and germanium photoconductive detectors

Host	Impurity	Cut-off wavelength(μm)
Si	In	8.0
Si	Ga	18
Si	Ni	5.4
Si	Mg	11
Ge	Au	9
Ge	Hg	14
Ge	Cu	27

but has now been replaced with mercury cadmium telluride $Hg_{1-x}(d_xT_e)$ which can operate between 2 and 14 μm. Such photoconductive detectors are marketed as single detectors and high density arrays with tailored long-wavelength cut-offs (Long and Schmidt 1970).

2.8 PHOTOVOLTAIC DETECTORS (PHOTODIODES)

Photodiodes act on the same general principles as photoconductors; that is, the generation of electrons and holes in a semiconductor by the absorption of light. However, the photodiode provides some internally generated electric field which can separate any electron–hole pairs that are photo-generated, giving rise to an electric current in an external circuit. There are several principal types of photodiode, the homojunction, the heterojunction, and the Schottky barrier featuring amongst them.

The homojunction diode is formed between two sections of the same semiconductor, one region being n-type, the other being p-type. At the interface region between the two regions, there is recombination of electrons and holes leaving a region devoid of free carriers. This region, known as the depletion region, contains ionized donors and acceptors and thus is a region of high field as well as of high resistance. The energy diagram for the p–n junction is shown in Fig. 2.20.

The depletion region acts as a barrier to current flow and is responsible for the well-known $I–V$ curve of the p-n diode (see curve marked 'dark' in Fig 2.21). When a forward bias is applied to the diode (i.e. a positive voltage applied to the p-side with respect to the n-side), the potential barrier is lowered, resulting in an increased flow of current. When a reverse bias is applied to the diode (a positive voltage applied to the n-side with respect to the p-side), the potential barrier increases. This increased barrier will inhibit the flow of holes from the p- to the n-region and of electrons from the n- to the p-region. However, there will, as a result of thermally generated electron–hole pairs, also be electrons in the p-type region and holes in the n-type region. These are known as minority carriers and, for them, the

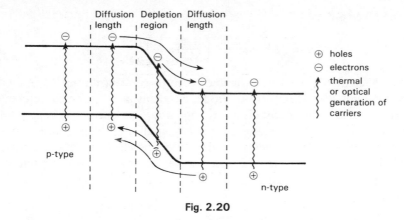

Fig. 2.20

potential barrier represents a drop in potential. Minority carriers generated in the depletion region will be swept across the junction producing a so-called reverse current. In addition, minority carriers produced outside the depletion region may diffuse into the high field region and be swept across the junction. However, only those carriers produced within the so-called diffusion length L of the depletion region, a measure of the length a minority carrier diffuses in its lifetime, will, on average, reach the depletion region, so contributing to the reverse current. For holes in germanium, $L = 2,000\ \mu$m.

The current that flows in the absence of light is (Kittel 1968)

$$I = I_0 \exp\left[\frac{eV}{kT} - 1\right]$$

where I_0 is the reverse current and V the voltage across the diode. In the case of reverse bias V_R, we can relate the width of the depletion region W

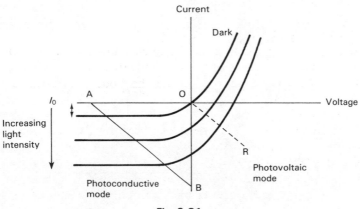

Fig. 2.21

to the reverse bias

$$W = \left[\frac{2V_R}{e} \left(\frac{1}{N_D} + \frac{1}{N_A} \right) \right]^{\frac{1}{2}}$$

where N_D and N_A are the donor and acceptor concentrations in the diode.

When light is incident on the diode, electrons and holes will be created by exactly the same mechanisms as with photoconductors. Those photo-carriers produced at (or within a diffusion length of) the p–n junction will be swept away by the depletion region field, the electrons into the n-type, the holes into the p-type (Fig. 2.20). This represents a negative current flow and displaces the I–V characteristics downward, as shown in Fig. 2.21. The current is now

$$I = I_0 \left[\exp\left(\frac{eV}{kT} - 1 \right) \right] - I_s$$

where I_s is directly related to the photon flux. The photovoltaic detector may be operated in several modes:

1. In the photvoltaic mode, we are operating in the fourth quadrant of Fig. 2.21 with no bias circuits. Figure 2.22 shows a generalized photovoltaic mode detector. When $R_L = \infty$, the open-circuit voltage is given by

$$V_{0c} = \frac{kT}{e} \ln\left(\frac{I_s + I_0}{I_0} \right)$$

As the light intensity increases, the open-circuit voltage increases logarithmically and we are moving along the positive V-axis. With a finite value of R_L, the photovoltaic detector would be operating along a line OR whose slope depended on the magnitude of R_L. When $R_L = 0$, it would be operating along the negative I-axis, and the short-circuit current produced would be

$$I_{sc} = -I_s$$

and so is linearly dependent on light intensity. To detect this current, the detector is connected to a transimpedance amplifier which transforms the short-circuit current into a voltage.

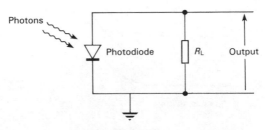

Photons

Photodiode R_L Output

Fig. 2.22

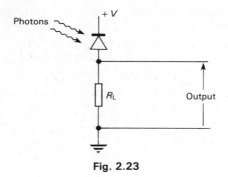

Fig. 2.23

2. In the photoconductive mode, the p-n junction is operated with reverse bias, as shown in Fig. 2.23. We are now operating in quadrant three in Fig. 2.21 (line AB). At high enough reverse voltage, a current $-(I_s + I_0)$ will flow. As the light intensity increases, the voltage across R_L increases proportionately. There are several advantages to backbiasing a photovoltaic detector. The reverse voltage extends the region of the depletion layer within the diode thus minimizing the capacitance of the diode. Since the electric field now extends through an appreciable part of the diode, photogenerated carriers travel with higher velocities and are gathered from a larger part of the diode than without bias. These factors enable the detector to respond better to weak and rapidly varying optical signals.

2.8.1 Some practical photodiode structures

Junction photodiodes: (a) p-n junction

A typical photodiode structure is shown in Fig. 2.24.

The photodiode consists of a highly doped thin p-type layer (the p^+ layer) on an n-type layer. The depletion region is maintained wholly in the n-region and its width depends on the reverse bias voltage. The antireflection coating, a thin film of silica or silicon nitride, reduces losses at the

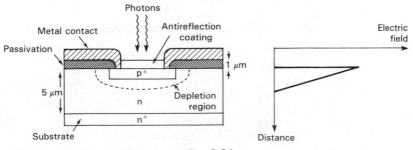

Fig. 2.24

air—semiconductor interface. The diode structure is grown on a heavily doped n^+ substrate to which electrical connection can be made. The diode structure shows front illumination. Back illumination may also be employed (Burrus *et al.* 1979). This configuration requires no photosensitive area for contacts, allowing the capacitance and material defects in the device to be minimized. It allows a relatively deep junction, since now the top layer need not be thin to reduce absorption loss. Photodiodes can also be employed using side illumination, resulting in considerable improvement in both quantum efficiency and transit times, since photons can then be absorbed entirely within the depletion region. This is especially relevant for detectors operating in spectral regions where the absorption is across an indirect gap so that the optical penetration depth may greatly exceed the optimum depletion layer width.

Junction photodiodes: (b) p-i-n diodes

The structure of a typical p-i-n photodiode is shown in Fig. 2.25.

The device shown is grown on a gallium arsenide substrate with a region of graded composition up to the absorbing region of gallium indium arsenide (Smith *et al.* 1980). The composition of the absorbing region can be adjusted to maximize the quantum efficiency at the operating wavelength). The weakly doped n^-GaInAs layer (the intrinsic or i-component in the p-i-n diode) is fully depleted at about 10 V reverse bias. The depletion region thus equals the width of the i-region and does not increase significantly with increased reverse bias. The device capacitance is thus fixed during manufacture. For a 100 μm diameter active area device, as shown above, a capacitance of 0.2 pF when fully depleted has been achieved. This device has a 50 percent quantum efficiency for a deviced optimized for 1.3 μm operation without an antireflection coating. The dark current of such narrow band p-i-n diodes has been observed to increase almost exponentially with voltage at high bias voltages (Fig. 2.26), due probably

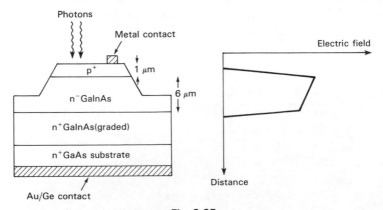

Fig. 2.25

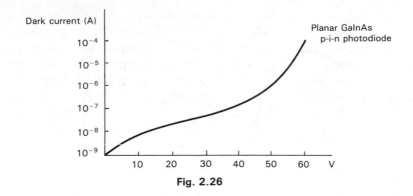

Fig. 2.26

to tunneling, and so, for low noise operation, should be operated at voltages well below breakdown.

Schottky photodiodes

When using a junction photodiode in a spectral region where the optical absorption coefficient is very high, the top p$^+$ layer needs to be very highly doped and very thin. Surface recombination of the photogenerated carriers can significantly reduce the quantum efficiency of the photodiode. This can be overcome by the use of the Schottky photodiode shown in Fig. 2.27.

The Schottky contact is an extremely thin (typically 10 nm) film of, for example, platinum, so that it is transparent to the radiation. A practical example of such a diode is that discussed by Wang (1983). It consists of a 25 μm diameter, 10 nm thick, semitransparent platinum film which forms the Schottky contact to an n$^-$ GaAs epitaxial layer (with electron concentration of 10^{15} cm^{-3}) grown on a n$^+$ GaAs substrate. With a 2 V reverse bias, a 1 μm wide depletion region is obtained. The junction capacitance has a value of around 100 fF at these reverse bias voltages and a very low reverse leakage current of about 6 pA.

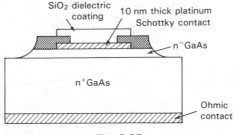

Fig. 2.27

2.8.2 Materials for photodiodes

Materials considerations for photodiodes are very much the same as for photoconductors. In the spectral region 0.8 to 1.6 μm (the important optical

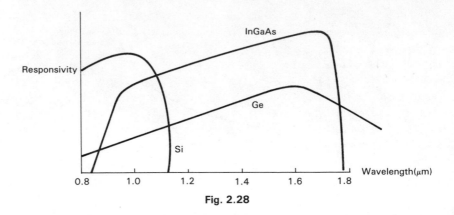

Fig. 2.28

communications band), three materials are currently dominant in photodiodes – silicon, germanium and indium gallium arsenide. The typical responsivities of these materials are shown in Fig. 2.28. Between 0.8 and 0.9 μm, silicon photodiodes are readily available and are of high quality due to the advanced state of silicon technology. In the 1.3 to 1.6 μm range, germanium photodiodes have been widely used. Germanium has, however, an indirect band gap, and relatively high leakage currents have hastened the development of alloys of elements from groups III and V of the periodic table, such as gallium indium arsenide which can be grown lattice-matched to substrates of indium phosphide (Lee *et al.* 1981). Further in the infrared, indium antimonide photodiodes are used in the 5 μm region while, around 3 μm, indium arsenide photodiodes are readily available.

2.8.3 Frequency response of photodiodes

When carriers are photogenerated in the diode, there are several effects which can limit its time response. It has been shown that carriers produced within the depletion region of the diode are separated by the internal field at the junction, while those created a diffusion length outside the depletion region must first diffuse to this region before separation can occur. These transport effects will limit the response time of the diode, in addition to the effects of the capacitance of the junction region (Godfrey 1979, Medved 1974, Wang 1983).

Fast photodiodes are usually designed to minimize the effects of the diffusion current. This is achieved by ensuring that the light penetrates the p-type region first and making the width of the depletion region sufficiently wide that most photoelectrons are generated in it. Photodiodes have p-type regions of less than 1 micron thick, thus minimizing the diffusion time τ_{diff}, where

$$\tau_{\text{diff}} = \frac{W_{\text{p}}^2}{2.43 D_{\text{n}}}$$

where W_p is the thickness of the p-region and D_n the minority carrier diffusion constant.

Carriers within the depletion region are now separated by the depletion field and drift across this region to their respective regions, the holes to the p-type layer, the electrons to the n-type. This takes a finite time and thus will also act to limit the frequency response of the photodiode. This drift time can be written

$$\tau_{drift} = \frac{W_I}{2.8v_s}$$

where W_I is the width of the depletion region, and v_s the saturated drift velocity of the carriers in the depletion region (typically about $5 \times 10^4 \, \mathrm{m\,s^{-1}}$).

The third limitation to the frequency response arises because there are a resistance and a capacitance associated with a diode. The capacitance arises because the redistribution of free charge carriers when the bias voltage changes, while there is a shunt resistance associated with the junction r_d and a series resistance associated with the n- and p-type layers. A simple model for these effects is shown in Fig. 2.29.

R_s would include the effect of any load resistance connected across the diode and, in general, $R_s \ll r_d$. Typically, R_s would be of the order of 50 Ω while r_d would depend on the mode of operation of the diode, being smaller in the photovoltaic mode (a few megohms) than in the photoconductive mode (tens of giga ohms) (Ross 1979).

The capacitance of the diode is given by

$$C_d = \frac{\varepsilon A}{W_I}$$

where A is the area of the diode and W_I the width of the depletion region (or the intrinsic region in a p-i-n diode).

In response to a pulse of light, this RC effect will cause the output of the photodiode to exhibit a rise time given by

$$\tau_{RC} = \frac{R_s C}{(1 + R_s/r_d)} = R_s C \qquad \text{for } R_s \ll r_d$$

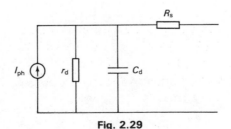

Fig. 2.29

The total rise time of the photodiode is given by

$$\tau = (\tau_{\text{diff}}^2 + \tau_{\text{drift}}^2 + \tau_{RC}^2)^{\frac{1}{2}}$$

$$= \left[\left(\frac{W_p^2}{2.43D_n} \right)^2 + \left(\frac{W_I}{2.8v_s} \right)^2 + \left(\frac{R_s \varepsilon A}{W_I} \right)^2 \right]^{\frac{1}{2}}$$

There are several observations to be made from this formula. The rise time will depend on the depletion width, W_I. Decreasing the width of the depletion region will reduce the drift time but increase the junction capacitance. The rise time is maximized with respect to W_I when

$$\tau_{\text{drift}} = \tau_{RC}$$

The value of the optical absorption coefficient determines whether drift or diffusion processes limit the frequency response of the diode. For direct band gap material with an optical absorption coefficient $\alpha > 10^4 \text{ cm}^{-1}$, most of the incident light will be absorbed in the front p-type layer. Diffussion of carriers to the junction will then limit the frequency response of the photodiode. For indirect band gap materials, such as silicon or germanium where $\alpha < 10^4 \text{ cm}^{-1}$, the junction thickness is made of the order of α^{-1} and the p-type region made extremely thin so that most of the light is absorbed in this region; drift across the junction now becomes the limiting factor. The junction region will be quite thick and such a photodiode is best implemented in a p-i-n structure. It will be necessary of course to check that, for the junction width, there is no RC degradation of the frequency response.

In practice there will be a trade-off between the thickness of the depletion region (and hence the quantum efficiency) and the speed of the device for indirect band gap materials. Wide depletion regions mean high quantum efficencies but will increase the drift time across the depletion region. For example, with silicon photodiodes at wavelengths of around 900 nm, high quantum efficiencies imply drift regions of about 50 μm giving the diode a response time of about a nanosecond. Faster speeds than this would mean a loss of quantum efficiency.

Consider the design of a germanium detector having a rise time of 50 ps. We will choose a front p-layer of 0.5 microns so that, with a diffusion constant $D = 10^{-2} \text{ m}^2\text{s}^{-1}$,

$$\tau_{\text{diff}} = \frac{(0.5 \times 10^{-6})^2}{2.43 \times 10^{-2}}$$

$$\approx 10 \text{ ps}$$

We will ignore this contribution. As shown previously, the depletion width needed to maximize the frequency response is

$$\tau_{\text{drift}} = \tau_{RC}$$

Since the total rise time is

$$\tau^2 = \tau^2_{\text{drift}} + \tau^2_{RC}$$

then, for maximum frequency response,

$$\tau = \sqrt{2} \cdot \tau_{\text{drift}} = \frac{\sqrt{2} \cdot W_I}{2.8 v_s}$$

For germanium, $v_s = 6 \times 10^4 \text{ m s}^{-1}$

$$W_I = 0.9 \times 6 \times 10^4 \times 50 \times 10^{-12}$$
$$= 3 \text{ microns}$$

Assuming that the diode works into a a 50 Ω load,

$$\tau_{RC} = R_L C = \frac{R_L \varepsilon A}{W_I} = \frac{50 \times 10^{-12}}{\sqrt{2}}$$

which, using a value of $\varepsilon = 16$, gives an area of photodiode of

$$A = 1.5 \times 10^{-5} \text{ cm}^2$$

The quantum efficiency of the device is

$$\eta = (1 - R)[1 - \exp(-\alpha W_I)]$$

and ignoring the surface reflectivity R,

$$\eta = 1 - \exp(-10^3 \times 3 \times 10^{-4})$$
$$= 1 - \exp(-0.3)$$
$$= 26\%$$

2.9 AVALANCHE PHOTODIODES (APDs)

It is possible to achieve internal current gain in a reverse-biased photodiode by means of a process known as impact ionization. Electrons and holes traversing the high electric field region of a semiconducting photodiode will gain kinetic energy, and, in a sufficiently high field, can acquire enough energy to create additional electron hole pairs via an inelastic collision process whereby lost energy is used to excite an electron from the valence band to the conduction band. This is called impact ionization. These newly created carriers, in turn, can create further carriers and so on, producing an avalanche of carriers. Thus the absorption of a photon will result in the liberation of a large number of secondary electrons, and the device is the solid state analog of the photomultiplier. Significant avalanche gain is obtained in semiconductors when the p-n junction is reverse biased to greater than half the breakdown voltage.

In most semiconductor materials, both electrons and holes can contribute to the impact ionization, which severely complicates any mathematical

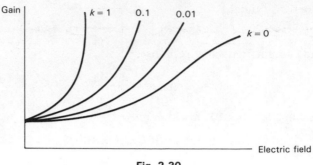

Fig. 2.30

treatment (see Webb *et al.* 1974). The gain as a function of the weighted average k of the ratio of the hole and electron ionization rates is shown in Fig. 2.30 for a 1 μm wide p-i-n diode.

For values of k other than zero, the gain of the device is a strong function of the electric field across the diode. Small variations in this field, caused by instabilities in power supplies or nonuniformities in doping levels in the semiconductor, can cause large fluctuations in gain. For example, at a gain of 100 with $k = 0.01$, a 0.05 percent fluctuation in electric field can give a 20 percent fluctuation in gain.

2.9.1 Frequency response of avalanche photodiodes

The frequency response of avalanche photodiodes is limited by two factors, the time taken for the carriers to drift to the high field region and the time for multiplication. In the so-called quasistatic regime (Kuvas and Lee 1970), the avalanche and drift regions can be treated separately. The total time constant of the avalanche photodiode will then simply be the sum of the transit time and multiplication time. In most avalanche photodiodes, the transit time limitation is the main factor. The frequency-dependent gain as a result of multiplication time effects can be written

$$M(\omega) = \frac{M(0)}{1 + j\omega\tau_m M(0)}$$

where $M(0)$ is the low frequency gain. τ_m is typically a picosecond or less and this effect does not become a limiting factor until the gain–bandwidth product of the photodiode is several hundred gigahertz.

2.9.2 Common avalanche photodiode structures

Figure 2.31 shows two common avalanche photodiode structures, the guard ring (a) and the reach-through (b).

The guard ring structure eliminates small areas with low breakdown

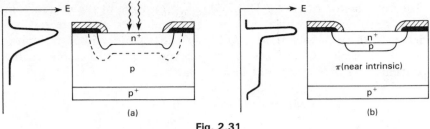

Fig. 2.31

voltages than the remainder of the junction and excessive leakages at junction edges. This ensures spatial uniformity of carrier multiplication over the entire light-sensitive area. The reach-through structure combines high speed, high gain and relatively low noise. When a reverse bias voltage is applied, the depletion layer of the diode (reach-through) to the low concentration π region when the electric field at the junction is about 10 percent less than required for avalanche breakdown. Addition of further applied voltage will cause the depletion layer to increase rapidly out to the p^+ contact although the field throughout the device increases only slowly. The diode is operated only in the fully depleted mode. Electrons generated in the π region are swept to the high field region where avalanche multiplication occurs. The resultant holes are then swept at almost saturation velocity to the p^+ contact.

2.10 NOISE IN PHOTODIODES

Since photogenerated carriers in a photodiode have to cross the depletion region, we may expect shot noise to be a fundamental mechanism for dark noise in these devices. The dark current is caused by the spontaneous creation of electron–hole pairs near the junction region. Thermal noise is also present due to the unavoidable circuitry associated with these devices. The photodiode is usually connected to some load resistance (for example, the input resistance of an amplifier) and, in the reverse bias configuration, the bias circuit itself will contribute to the thermal noise. Finally, $1/f$ noise will also exist due to various causes, such as electronic transition involving surface states or by electrical contacts to the detector.

2.10.1 Noise in APDs

The APD will contain the usual noise terms with one added complication. The process of carrier multiplication by a factor M increases the shot noise fluctuations of the photocurrent $\langle i_{ph} \rangle$ to give the expression

$$\langle i_n^2 \rangle = 2e\langle M^2 i_{ph} \rangle \, \Delta f$$
$$= 2e\langle i_{ph} \rangle \langle M \rangle^2 F(M) \, \Delta f$$

The noise factor $F(M) = \langle M^2 \rangle / \langle M \rangle^2$ is a measure of the degradation of the signal compared with an ideal noiseless multiplier, and increases considerably with average gain $\langle M \rangle$. $F(M)$ depends on the ratio of the ionization rates of electrons and holes and on the type of carrier that initiates the avalanche. McIntyre (1966) has shown that

$$F(M) = k_w M + (1 - k_w)\left(2 - \frac{1}{M}\right)$$

k_w is related to k, the ratio of ionization rates for holes and electrons, by

$$k_w = k \text{ for an electron-initiated avalanche}$$

$$= \frac{1}{k} \text{ for a hole-initiated avalanche}$$

If the ionization rates are equal, the excess noise factor increases as the carrier multiplication rate, and is at its maximum. For this reason, relatively high excess noise can be expected in germanium APDs. Much lower noise is observed in junctions with highly unequal ionization rates if the avalanche is initiated with carriers of higher ionization rate, the excess noise factor $F(M)$ tending towards a limit of 2 at high multiplications. Silicon is, for this reason, one of the most suitable materials for APDs. However, beyond about 1.1 μm, germanium is presently the best available material in the region 1.1 μm to 1.4 μm, especially for fibre optic communication systems operating at high bit rates, where the amplifier noise tends to become more of a limiting factor. Development of APDs in materials such as $In_xGa_{1-x}As$, $GaAs_{1-x}Sb_x$, and $Hg_{1-x}Cd_xTe$ seem likely to supersede germanium in this region of the spectrum.

In addition to the above noise multiplication, the bulk dark current is also subject to the avalanche process with consequent excess multiplication noise.

2.10.2 Minimum detectable signal in photodiodes

In this section, we shall develop a model of the photodiode which will enable us to calculate the minimum detectable signal of the photodiode. We have seen that, in the presence of an optical signal of intensity I_s, the current in the diode can be written

$$\langle I \rangle = \underbrace{I_0\left[\exp\left(\frac{eV}{kT}\right) - 1\right]}_{\text{Dark current}} \underbrace{- \langle I_{ph} \rangle}_{\text{Photocurrent}}$$

The first term is the dark current in the diode, while the second is the photon-induced current. The dark current is the difference of two currents:

a forward current $I_0 \exp(eV/kT)$, due to majority carrier current flow across the junction; and reverse current I_0, due to minority carrier flow in the reverse direction. These two currents are statistically independent. Thus, the shot noise of the total current flowing in the photodiode is

$$\langle i^2 \rangle = 2e\,\Delta f[I_0 \exp(eV/kT) + I_0 + \langle I_{\text{ph}} \rangle \rangle$$

where $\langle I_{\text{ph}} \rangle$ is the photocurrent.

For the case of a reverse-biased diode (photocurrent mode)

$$\exp(eV/kT) \to 0$$

and

$$\langle i^2 \rangle = 2eI_0\,\Delta f + 2e\langle I_{\text{ph}} \rangle\,\Delta f$$

which is the sum of the shot noise due to the reverse current and that due to the photocurrent.

For the photodiode acting in the photovoltaic mode

$$\exp(eV/kT) \to 1$$

and

$$\langle i^2 \rangle = 2e\,\Delta f[I_0 + I_0] + 2e\langle I_{\text{ph}} \rangle\,\Delta f$$
$$= 4eI_0\,\Delta f + 2e\langle I_{\text{ph}} \rangle\,\Delta f$$

The incremental resistance of the diode is

$$r_{\text{d}} = \left[\frac{\partial V}{\partial I}\right] \text{ dark}$$

$$= \frac{kT}{eI_0} \text{ at } V = 0$$

In this case, the mean square noise current is

$$\langle i^2 \rangle = \frac{4ekT}{er_{\text{d}}}\,\Delta f + 2e[I_{\text{ph}} \rangle\,\Delta f$$

$$= \frac{4kT}{r_{\text{d}}}\,\Delta f + 2e\langle I_{\text{ph}} \rangle\,\Delta f$$

The dark noise in the photovoltaic mode is simply the thermal noise of the incremental resistance of the diode.

We are now in a position to derive a model for the photodiode in the photocurrent mode. We shall leave the photovoltaic mode detector until the chapter on amplifiers, since, in this mode, the photodiode is usually employed with a transimpedance amplifier.

The general arrangement of the photodiode is shown in Fig. 2.32. R_{B} is a biasing resistor, while R_{L} is the load across which the photon-induced voltage is developed. In terms of the general noise model for photodiodes

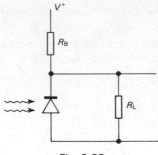

Fig. 2.32

developed in the section on photomultipliers, $G = 1$ and $F(G) = 1$. The model of the photodiode is shown in Figure 2.33.

$\langle i_s^2 \rangle$ is the shot noise due to the photons. These photons will come from the signal of average optical power P_{ph} and background photons with average optical power P_B. The mean photocurrent is

$$\langle I_s \rangle = \langle I_{ph} \rangle + \langle I_B \rangle = \frac{e\eta}{hf_s} (P_{ph} + P_B)$$

$\langle i_d^2 \rangle$ is the shot noise due to the dark current. $R_\parallel$ is the parallel combination of R_L and R_B with shot noise current

$$\langle i_R^2 \rangle = \frac{4kT}{R_\parallel} \Delta f$$

$$\frac{S}{N} = \frac{\left(\frac{e\eta}{hf_s}\right)^2 P_{ph}^2}{2e\,\Delta f(\langle I_{ph} \rangle + \langle I_B \rangle + \langle I_d \rangle) + 4kT\,\Delta f/R_\parallel}$$

In most practical cases, the need to make the bandwidth due to RC effects as large as possible forces as low a value of R_L as possible, making the thermal noise due to R_L dominate the noise. In this case

$$\frac{S}{N} = \frac{\left(\frac{e\eta}{hf_s}\right)^2 (P_{ph})^2}{4kT\,\Delta f/R_L}$$

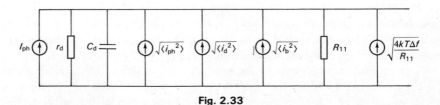

Fig. 2.33

giving a minimum detectable power

$$(P_{ph})_{min} = \left(\frac{4kT\,\Delta f}{R_L}\right)^{1/2} \frac{hf_s}{e\eta}$$

If the shot noise of the signal dominates the noise

$$\frac{S}{N} = \frac{(e\eta/hf_s)^2(P_{ph})^2}{2e\,\Delta f\langle I_{ph}\rangle}$$

Now

$$\langle I_{ph}\rangle = \left(\frac{e\eta}{hf_s}\right)\cdot P_{ph}$$

so that

$$\frac{S}{N} = \frac{\left(\frac{e\eta}{hf_s}\right)^2\cdot(P_{ph})^2}{2e\,\Delta f\cdot\left(\frac{e\eta}{hf_s}\right)P_{ph}}$$

giving a minimum detectable power

$$(P_{ph})_{min} = \frac{2hf_s\,\Delta f}{\eta}$$

Often in using photodiodes, the photon flux incident on the diode is modulated in some way. Let us suppose that it is intensity modulated:

$$P_{ph} = P_0(1 + m\cos\omega t)$$

The important parameter is now the minimum depth of modulation m which can be detected by the the photodiode. In this case, the 'signal' is the component of the photocurrent at the frequency ω and has an r.m.s. value of

$$\langle I^2_{r.m.s}\rangle = \left(\frac{e\eta}{hf_s}\right)^2\cdot\tfrac{1}{2}m^2P_0^2$$

The signal-to-noise ratio is now

$$\frac{S}{N} = \frac{(e\eta/hf_s)^2\cdot\tfrac{1}{2}m^2P_0^2}{2e\,\Delta f(\langle I_{ph}\rangle + \langle I_B\rangle + \langle I_d\rangle) + 4kT/R_{\parallel}\cdot\Delta f}$$

and the minimum depth of modulation which can be detected occurs when this is equal to 1. Thus, for example, when thermal noise is the dominant noise mechanism, the minimum observable depth of modulation is

$$(mP_0)_{min} = \left(\frac{hf_s}{e\eta}\right)\cdot\left(\frac{8kT\,\Delta f}{R_L}\right)^{1/2}$$

2.10.3 Minimum detectable signal in APDs

The minimum detectable optical power in an avalanche photodiode is easily obtained. The model is exactly the same as for the photomultiplier (using M in place of G, as is the convention for avalanche photodiodes). Again, we will assume that the optical power is intensity modulated. However, as discussed previously, the excess noise factor is quite complicated. We will assume for simplicity that $F(M) = M^x$ where $0 < x < 1$. The signal-to-noise power ratio is

$$\frac{S}{N} = \frac{\dfrac{m^2}{2}(e\eta P_0/hf_s)^2 M^2}{\left\{ 2M^{2+x} e\left[\langle I_d \rangle + \dfrac{\eta e}{hf_s}(P_0 + P_B) \right] \Delta f + 4\dfrac{kT\,\Delta f}{R_L} \right\}}$$

Under the conditions where the thermal noise dominates the shot noise

$$\frac{S}{N} = \frac{m^2}{2}\left(\frac{e\eta P_0}{hf_s}\right)^2 M^2 \frac{R_L}{4kT\,\Delta f}$$

which increases as M increases. Eventually, as M gets larger and larger, the shot noise term will begin to dominate, causing the signal-to-noise ratio to fall as M^{-x}. It is a simple matter to show that the signal-to-noise is maximized as a function of the multiplication gain M when

$$2M^{2+x} e\left\{ \langle I_d \rangle + \frac{\eta e}{hf}(P_0 + P_B) \right\} \Delta f = \frac{2}{x}\cdot\frac{4kT\,\Delta f}{R_L}$$

and the minimum power detectable is

$$(mP_0)_{\min} = \frac{hf_s}{e\eta M}\left[\frac{2kT\,\Delta f}{R_L}\cdot\left(1 + \frac{2}{x}\right) \right]^{\frac12}$$

showing that the avalanche photodiode has greater detection sensitivity than the photodiode (M typically is of the order of ~ 100).

2.11 NOISE AND MINIMUM DETECTABLE SIGNAL IN PHOTOCONDUCTORS

There are two sources of noise in photoconductors. The first is generation–recombination noise, caused by the statistical fluctuations in both generation and recombination rates which initiate and terminate the carrier existence. We should notice an important difference between the noise associated with dark generation of charge in photodiodes and photoconductors. This is due to the substantial lifetime of carriers in the photoconductor. In the photodiode, the carriers are rapidly swept apart and out of the depletion region. Their *effective* lifetime is very short and only the statistical fluctuations in their generation is important. In photoconductors, however, the carriers have substantial lifetime and there are now two random processes contributing to the noise-generation and recombination.

The generation–recombination noise can be written in the form (van Vliet 1958):

$$\langle i_{GR}^2 \rangle = 4e\left(\frac{\tau}{T}\right)\frac{\langle I \rangle \, \Delta f}{1 + \omega^2 \tau^2}$$

where

$$\langle I \rangle = \frac{\tau}{T}(\langle I_{ph} \rangle + \langle I_B \rangle)$$

is the mean d.c. current from the photoconductor, being due to the signal and background photons. To bring this into the form of the standard model, this can be written

$$\langle i_{GR}^2 \rangle = 2e(\langle I_{ph} \rangle + \langle I_B \rangle)G^2(\omega)F \, \Delta f$$

where

$$G(\omega) = \frac{G(0)}{(1 + \omega^2 \tau^2)^{1/2}} = \frac{\tau/T}{(1 + \omega^2 \tau^2)^{1/2}}$$

and $F = 2$.

The factor of 2 represents the fact that both generation and recombination are random, uncorrelated events.

Thermal noise arises from two sources: noise associated with the bias circuits of the photoconductor, and noise associated with the resistance of the photoconductor itself. $1/f$ noise will, of course, also be present to varying degrees.

The model of the photoconductor is shown in Fig. 2.34. In this model, G_0 represents the conductance of the photoconductor and G_L that of the load, C represents the capacitance of the photoconductor and its associated output circuitry, and $\langle i^2 \rangle$ is the sum of the signal and background fluctuations. The signal-to-noise power is thus

$$\frac{S}{N} = \frac{\left(\dfrac{\eta e}{hf_s}\right)^2 P_{ph}^2 \cdot \left(\dfrac{\tau}{T}\right)^2}{2e(\langle I_{ph} \rangle + \langle I_B \rangle)G(\omega)F \, \Delta f + 4kT(G_0 + G_L)\, \Delta f}$$

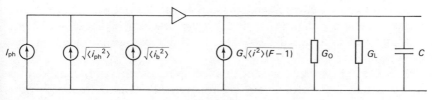

Fig. 2.34

In this situation, where generation–recombination noise is the dominant mechanism,

$$\frac{S}{N} \approx \frac{\left(\frac{\eta e}{h f_s}\right)^2 \cdot P_{ph}^2}{2e\langle I_{ph}\rangle \tau/T 2\Delta f} = \frac{\left(\frac{\eta e}{h f_s}\right) \cdot P_{ph}\left(\frac{\tau}{T}\right)^2}{2e(\tau/T)^2 \, 2\Delta f}$$

so that

$$(P_{ph})_{min} = \frac{4\,h f_s \,\Delta f}{\eta}$$

2.12 THERMAL DETECTORS

Optical radiation absorbed by a thermal detector causes its temperature to rise, and this temperature change causes a change in some property of the detector which can be detected. In the ideal thermal detector, the optical energy is absorbed uniformly at all wavelengths. In practice, absorption by the sensing layer is wavelength dependent, and the detecting elements are usually blackened to produce an approximately wavelength independent response (Huang 1978, de Waard and Wormser 1959).

The thermal detector can be considered as a thermal capacity C connected to a heat sink at temperature T via a thermal conductance G. If the detector is subjected to a time-varying energy flux $W(t)$ so that its temperature is $T + \Delta T$, then

$$W(t) = G\,\Delta T + C\frac{\mathrm{d}}{\mathrm{d}t}\Delta T$$

If the radiation flux is suddenly removed, ΔT will decay exponentially with time constant

$$\tau = \frac{C}{G}$$

We can thus see the factors which affect the speed of response of thermal detectors. When operated at room temperature (their normal mode of operation), these detectors are generally slow (time constant of milliseconds to a few seconds) although cooling, which reduces the thermal capacity of the detecting element, can lead to an improvement in these figures. We can speed up the response by increasing the thermal contact with the heat sink, but this will also prevent the thermal detector reaching higher temperatures. Thus, there is a trade-off between sensitivity and response in thermal detectors.

These detectors work by sensing changes in temperature, and it should be noted that they are useful chiefly for modulated or pulsed radiation only.

Some of the major types of thermal detector will now be discussed.

2.12.1 Thermopile

This is one of the oldest infrared detectors. It is essentially several thermo-couple junctions arranged in series. The basic element is a junction between two dissimilar conductors having a large Seebeck coefficient. The active (hot) junction is deposited on a free-standing film and is blackened. This makes the system an efficient absorber with low thermal mass. The cold junction is also deposited on the same film but over a heat sink. To perform efficiently, a large electrical conductivity is required to minimize joule heat loss and a small thermal conductivity to minimize heat conduction loss between hot and cold junctions. These requirements are not compatible and design compromise has to be made, materials such as Bi_2Te_3 and related compounds being popular.

Modern thin-film thermopiles have several advantages. They are highly reliable and rugged with long term stability. They require no bias voltage and can be used to detect modulated or direct radiation. Their main draw-backs are the slow response time and insensitivity compared with other thermal detectors.

2.12.2 Bolometer

The bolometer is a resistive element constructed from a material with a large temperature coefficient of resistance; absorbed radiation produces a large change in resistance. Bolometers are made out of metal film and out of semiconductors (thermistor). Semiconductor bolometers have a large coefficient of resistance especially at very low temperatures ($4\%/^\circ C$) which is an order of magnitude larger than metals. Such semiconductor bolometers are especially useful at very low temperatures since they can combine high sensitivity with fast response. Doped silicon or germanium operated at liquid helium temperatures can have, over most of the far infrared, a uniform performance comparable in sensitivity with the best photon detectors.

To detect the resistance change in a flake of semiconductor, a steady bias voltage is applied across the detector. Changes in ambient temperature can cause drift, and to minimize this the bolometer is constructed of two flakes of semiconductor connected in series. The active flake is coated in black while the other (or compensating) flake is shielded from the radiation.

2.12.3 Pyroelectric detectors

A pyroelectric material has a low enough crystalline symmetry to possess a permanent internal electric dipole moment, and will exhibit an internal electric field along a certain crystal axis. If conducting parallel electrodes are applied to the crystal, charge will be stored in these electrodes. A change of temperature will change the lattice spacing of the material, and hence the internal electric field. The change in stored charge on the electrodes can be detected as a current in external circuitry.

The magnitude of the pyroelectric effect is such that the sensitivity of the best detectors is comparable with that of Golay cells or thermopiles. It is relatively fast (response times can be as short as microseconds) but still much slower than photon detectors.

The material in general use for pyroelectric detectors is triglycerine sulphate (TGS) although materials such as lithium tantalate and polyvinyl fluoride (PVF) are also used.

2.12.4 Dark noise in thermal detectors

In the absence of electromagnetic radiation, noise in thermal detectors will be due to the usual factors — Johnson noise associated with the resistance of the element, temperature noise due to fluctuations in the element temperature, and $1/f$ noise associated with contacts at the element.

Minimum detectable power in a bolometer in the infrared

To operate a bolometer, an accurately controlled bias current i from a suitable source is passed through the element.

If suitably modulated radiation is incident on the bolometer and produces an increase in temperature δT, then the bolometer will change its resistance by

$$\delta R = \alpha R \delta T$$

where α is the temperature coefficient of resistance of the bolometer.

The output voltage thus generated is

$$V_S = i\delta R = i\alpha R \delta T$$

If the thermal conductance of the bolometer is G and its thermal capacitance is C, then we can relate δT to the instantaneous incident power

$$P(t) = G\delta T + C \frac{\mathrm{d}}{\mathrm{d}t} \delta T$$

For

$$P(t) = P_0 \, \mathrm{e}^{j\omega t}$$

$$\delta T = \frac{P_0 \, \mathrm{e}^{j\omega t}}{G + j\omega C}$$

Thus, the amplitude of the signal voltage generated is

$$V_s = \frac{i\alpha R P_0}{(G^2 + \omega^2 C^2)^{1/2}}$$

There are two main sources of noise. The first is Johnson noise in the resistive element:

$$\langle v_{n1}{}^2 \rangle = 4kTR\,\Delta f$$

The second is due to temperature fluctuations from the background radiation with mean square power

$$\langle \Delta W^2 \rangle = 16A\,\varepsilon\sigma kT^5\,\Delta f$$

and hence a mean square voltage fluctuation

$$\langle v_{n2}{}^2 \rangle = \frac{i^2\alpha^2 R^2}{(G^2 + \omega^2 C^2)} \cdot 16A\varepsilon\sigma kT^5\,\Delta f$$

Finally, there will be noise associated with the input impedance of the associated amplifier:

$$\langle v_{n3}{}^2 \rangle = 4kT\,R_{\text{in}}\,\Delta f$$

Thus, the signal-to-noise power ratio is

$$\frac{S}{N} = \frac{i^2\alpha^2 R^2 P^2/(G^2 + \omega^2 C^2)}{\left\{4kTR + \dfrac{i^2\alpha^2 r^2}{(G^2 + \omega^2 C^2)} \cdot 16A\varepsilon\sigma kT^5 + 4kTR_{\text{in}}\right\}\Delta f}$$

For a bolometer at high frequencies, the amplifier noise can be the ultimate limitation to the minimum power observable

$$\frac{S}{N} \approx \frac{i^2\alpha^2 R^2 P^2/\omega^2 C^2}{4kTR_{\text{in}}\,\Delta f}$$

and the minimum detectable power, when $\dfrac{S}{N} = 1$, is

$$P_{\text{min}} = \frac{\omega C}{i\alpha R}\,(4kTR_{\text{in}}\,\Delta f)^{1/2}$$

2.13 MICROCHANNEL PLATES

At the beginning of this chapter we saw that the photomultiplier was one of the fastest and most sensitive devices for detection of visible photons. In this discussion it was seen how the statistical variation in the conversion of photons to photoelectrons and the statistical variation in the secondary emission limit both the pulse height and time resolution of the device. We shall now discuss a relatively new device which is increasing in popularity and which is faster, less noisy, and more efficient than conventional photomultipliers (Leskovar 1972, 1984).

A schematic diagram of a microchannel plate is shown in Fig. 2.35. It consists essentially of a two-dimensional array of thousands of short single-channel electron multipliers of very small diameter, closely packed together.

Fig. 2.35

The electron multiplying channel is a glass tube of ~2 mm in length and 15–50 μm in diameter which has been coated on its inside surface with a high resistance semiconductor capable of emitting secondary electrons. The array of microchannels is collected electrically in parallel by metal electrodes on opposite faces of the plate. A typical voltage of 1 kV is maintained across the faces of the plate. A continuous potential gradient will be maintained along each microchannel by the semiconductor coating. When a photon strikes the inside of one of the microchannels, it releases a photoelectron which is accelerated along the channel. When it strikes the semiconductor again, it releases several secondary electrons which are further accelerated down the channel in an avalanche process until a burst of electrons is released at the output which may be collected at an anode. With the parameters given above, gains of 10^3 to 10^4 are possible. The cross-section of a single microchannel is shown in Figure 2.36.

Because of the small channel length and the high electric field within the channel, the electron transit time and its spread are very small, typically less than 10^{-9} seconds and 10^{-10} seconds respectively.

The semiconductor coating on the microchannel surfaces acts not only as a source of secondary electrons but also as a source of potential gradient. It acts thus as not only the equivalent of the dynodes but also the dynode resistor chain in a photomultiplier.

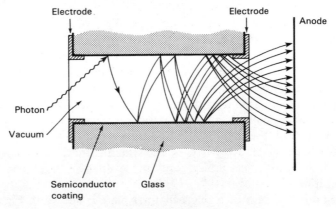

Fig. 2.36

2.13.1 Limits to the microchannel device

The simple straight microchannel plate device is limited to a maximum gain of $\sim 10^4$ primarily by ionization of residual gases inside the microchannels and by the ejection of atoms absorbed on the channel walls. The positive ions thus produced are accelerated towards the input end of the microchannels, where they may have gained sufficient energy to eject secondary electrons, thus producing spurious noise pulses.

The gain limitation is lifted to some extent by design modification of the device. Two or more microchannel plates are assembled such that the channels of one plate are at an angle to the channels of the next plate (the so-called chevron configuration). The electrons travel down the channels without being significantly inhibited by the angle. Positive ions travelling down the tube will travel only a short distance before striking the wall of the tube. They will not have gained sufficient energy to eject a significant number of secondary electrons. Successful high gain microchannel plate devices manufactured recently have gains of greater than 10^6. They consist of three microchannel plates in the so-called Z-configuration, where the channels in each plate are at an angle of $5°$ to $8°$ with respect to the plate surface normals. With this configuration, gains of the order of 10^7 can be achieved.

At sufficiently large microchannel plate voltage, the gain of a microchannel plate will saturate due to one of two causes. When a photon strikes a channel and secondary electrons are emitted, positive charges are left at the surface of the channel wall, which will increase in magnitude towards the output end. This charge will obviously repel electrons unless neutralized by current flow in the channel from the external power supply. Since the channel has quite a high resistivity, this can take up to a millisecond to achieve and such gain saturation will be important in the pulse mode of operation. In the continuous mode, another mechanism limits the gain, namely the limited current flowing in the channel wall. As more and more secondary electrons are emitted towards the output end, the wall current becomes increasingly depleted, reducing the electric field in the channel. This reduces secondary emission and hence the gain.

Operation in the saturation mode can have its advantages. It has been shown by several workers that, in this mode, fluctuations in gain from one pulse to another within a channel are much reduced over the nonsaturated mode. Also, the gain fluctuations from microchannel to microchannel within a device are reduced.

2.13.2 Microchannel plates and radiation

So far, we have not specified the type of photon which the microchannel plate is used to detect. In fact, the device, with suitable modifications can

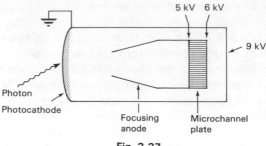

Fig. 2.37

be used to detect photons from soft X-rays through the ultraviolet to the infrared (it can also be used as a particle detector in nuclear physics). For soft X-rays and for ultraviolet, the microchannel plate used in a windowless configuration. For near ultraviolet thru infrared, the detection efficiencies of microchannel plates are too small for practical use and they must be used in conjunction with photocathodes, where they make excellent photon detectors with all the advantages discussed above. A typical arrangement for visible photons is shown in Fig. 2.37.

Photomultipliers such as this have been constructed using Z-channel and curved channel plates. They have gains of the order of 10^6 and photocathodes of 15 mm to 20 mm in diameter. They have electron transit times of about 0.6 ns with single photoelectron spreads of less than 100 ps. They have high immunity to magnetic fields, with axial fields of the order of a kilogauss not affecting the gain while needing transverse fields of the order of half a kilogauss to reduce the gain by half. This is an enormous improvement over conventional photomultipliers, with fields of the order of a gauss causing severe problems.

SUMMARY

In this chapter we have investigated the properties of some of the more important transducers of light, or, to give them their more usual title, photodetectors. We have broadly classified photodetectors as thermal detectors or photon detectors depending on the way in which the detector responds to the optical radiation incident upon it. In order to compare the performance of these detectors, several important parameters were discussed. These were spectral response, speed of response, and sensitivity. The first of these factors is roughly independent of wavelength for thermal detectors but is strongly wavelength dependent for photon detectors. The speed of response of a detector determines its ability to respond to a time-dependent optical source of power. Thermal detectors have, in general, a much slower speed of response than photon detectors. The final property, that of sensitivity, is the ability of the detector to measure weak optical signals in the presence of noise. A general model of a photodetector was developed showing the principal signal and noise sources. This model allowed calculations of signal-to-noise ratios and minimum detectable powers to be made for the photodetectors.

Although the three factors were discussed under separate headings, the discussion emphasized that they are interrelated and that trade-offs between these factors have to be made in any practical application.

REFERENCES

Burrow, C.A., Dentai, A.G. and Lee, T.P. 1979. InGaAsP p-i-n photodiodes with low dark current and small capacitance. *Electron Letters*, Vol. 15, pp. 655–6.

Godfrey, L.A. 1979. Designing for the fastest response ever. *Optical Spectra*, October, pp. 43–6.

Huang, C. 1978. Infrared thermal detectors: a comparison. *Optical Spectra*, September, pp. 47–50.

Jones, R.C. 1959. Noise in radiation detectors. *Proceedings of the IRE*, Vol. 47, pp. 1481–6.

Kittel, C.A. 1968. *Introduction to Solid State Physics*. New York: Wiley.

Kuvas, R. and Lee, C.A. 1970. Quasistatic approximation for semiconductor avalanches. *Journal of Applied Physics*. Vol. 41, pp. 1743–55.

Lee, C.A., Logan, R.A., Batdorf, R.L., Kleinmack J.J. and Wiegmann, W. 1964. Ionisation rates of electrons and holes in silicon. *Physical Review*. Vol. 134A, pp. 761–73.

Leskovar, B. 1972. Microchannel plates. *Physics Today*, November, pp. 42–9.

Leskovar, B. 1984. Recent advances in high speed photon detectors. *Laser Focus*, February, pp. 73–8.

Long, D. and Schmidt, J.L. 1970. Cadmium mercury telluride and closely related alloys. In *Semiconductors and Semimetals*, Vol. 5, Willardson, R. and Beer, A. (editors). New York: Academic Press.

McIntyre, R.J. 1966. Multiplication noise in uniform avalanche diodes. *IEEE. Transactions on Electronic Devices.*, Vol. 13, pp. 164–8.

Medved, D. B. 1974. Photodiodes for fast receivers. *Laser Focus*, January, pp. 99–106.

Oliver, B.M. 1965. Thermal and quantum noise. *Proceedings of the IEEE*, Vol. 53, pp. 436–54.

Putley, E.H. 1966. Solid state devices for infrared detectors. *Journal of Scientific Instruments*, Vol. 43, pp. 857–68.

Putley, E.H. 1973. Modern infrared detectors. *Physics in Technology*, Vol. 4, pp. 202–22.

Rose, A. 1955. Performance of photoconductors. Proceedings of the IRE, Vol. 43, pp. 1850–69.

Ross, D.A. 1979. *Optoelectronic Devices and Optical Imaging Techniques*, Macmillan.

Schultz, M.L. and Morton, G.A. 1955. Photoconduction in germanium and silicon. *Proceedings of the IRE*, Vol. 43, pp. 1819–29.

Sharpe, J. 1964. Dark current in photomultipliers. EMI symposium.

Smith, R.A. 1959. *Semiconductors*. Cambridge University Press.

van Vliet, K.M. 1958. Noise in semiconductors and photoconductors. *Proceedings of the IRE*, Vol. 46, pp. 1004–18.

de Ward, and Wormser, E.M. 1959. Description and properties of various thermal detectors. *Proceedings of the IRE*, Vol. 47, pp. 1508–13.

Wang, S.Y. 1983. Ultra-high speed photodiode. *Laser Focus*, December, pp. 99–106.

Webb, P.P., McIntyre R.J. and Conradi, J. 1974. Properties of avalanche photodiodes. *RCA Review*, Vol. 35, pp. 234–78.

PROBLEMS

2.1 A photomultiplier is irradiated with light of wavelength 600 nm. The photomultiplier has a gain of 10^6, a quantum efficiency of 10 percent, a dark current of 10^{-15} A, and an anode current rise time of 10 ns. The anode is connected to a load resistor of 10 MΩ and stray capacitance of 5 pF. Calculate the minimum detectable power of the photomultiplier.

2.2 A photodiode is illuminated with 1 μW of light at a wavelength of 500 nm. Calculate the resulting photocurrent if the active region of the device is 1 μm thick and the optical absorption coefficient is 10^4 cm^{-1} at this wavelength. (Ignore the reflectivity of the surface of the semiconductor.)

2.3 Light of wavelength 1 μm is incident on a photodetector of quantum efficiency 20 percent. If the arrival of photons at the detector follows a Poisson distribution, calculate the probability that no photoelectron will be emitted in a 1 μs interval if the incident optical power is 8×10^{-12} J.

2.4 The leakage current in a GaAs photodiode at low bias can be written

$$I_R = \frac{e n_i A W}{\tau_e}$$

where A is the device area, W the depletion region width, n_i the intrinsic carrier concentration and τ_e an effective thermal generation lifetime for carriers in the depletion region. Calculate I_R for a 6 mm diameter circular photodiode with $W = 4.5$ m at a reverse bias of 5 V at (a) $T = 300$ K and (b) $T = 77$ K. (The energy gap for GaAs is $E_g = 1.4$ eV and $\tau_e = 14$ ns.)

    Calculate the quantum efficiency for the device assuming a reflectivity of 10 percent and an absorption coefficient of $\alpha = 10^4$ cm^{-1}. Hence, calculate the shot noise of the photodiode per unit bandwidth at the two temperatures if the incident light intensity is 10^{-10} W.

2.5 A photodiode is constructed out of GaAs having a density of donors $N_D = 2 \times 10^{15}$ cm in the n-layer. Calculate the capacitance of the diode, assuming that the doping in the n-layer is very much less than that in the p-layer. Calculate the *RC* time constant of this diode assuming a 50 Ω load resistance ($\varepsilon_r = 11.5$ for GaAs).

2.6 Calculate D and D^* for a detector having a noise equivalent power of 5×10^{-10} W, an active area of 0.5 cm^2 and a noise bandwidth of 1 kHz.

2.7 Consider an avalanche photodiode with an active region of thickness W with constant electron and hole multiplication factors α and β respectively (see figure below).

    Making no assumption about which carrier type initiates avalanche, obtain

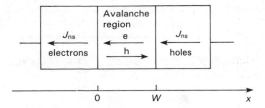

an expression for the multiplication M, where

$$M = \frac{(J_p + J_n)}{J_s}$$

$J_p(J_n)$ being the hole (electron) current in the avalanche region, and J_s the total (saturation) current flowing in an external circuit. Fot the case of an electron-initiated avalanche only, show that your expression reduces to

$$1 - (1/M) = \int_0^W \alpha \, \exp\left[-\int_0^x (\alpha - \beta) \, dx'\right] dx$$

2.8 A thermal detector has a thermal conductance G and a heat capacity C. Show that if there are random fluctuations in the power exchanged between the body and its surroundings, $W(t)$ (e.g. due to background noise), then

$$W(t) = G \, \Delta T + C \frac{d}{dt} \Delta T$$

where ΔT is the difference in temperature between the detector and its surroundings.

A pyroelectric detector is used to detect infrared radiation which is sinusoidally modulated at an angular frequency ψ. This causes the temperature of the detector to change by ΔT, and an alternating charge on the detector electrodes to appear, of magnitude

$$\Delta Q = pA \, \Delta T$$

where p is the pyroelectric coefficient and A the active area of the device. Obtain an expression for the r.m.s. amplitude of the signal current.

The pyroelectric detector behaves as a capacitor with a resistance R across it. Obtain an expression for the r.m.s. voltage across the electrodes of the pyroelectric element. Assuming thermal noise in R is the dominant noise source, derive an expression for the minimum detectable power.

3

Parallel Detection and Imaging of Light

So far, we have been concerned with the detection of light at essentially a single wavelength and at a single point in space. For example, suppose we are detecting light which is dispersed by a diffraction grating. We would place a slit in the path of the diffracted light so that essentially only one wavelength would pass through the slit, that wavelength being determined by the angular position of the grating. The detector is then placed behind the slit and, by rotating (or scanning) the grating, a readout of the intensity of light as a function of wavelength can be obtained (Fig. 3.1).

There are occasions, however, when this type of detection process is inappropriate and sometimes positively wasteful. We may, for example, be interested in the spatial variation in light intensity across a laser beam. In the case of spectroscopy, the conventional arrangement above only uses a fraction of the total available light at any one time. For a simple spectrometer scanning across a spectrum of, say, 500 nm with a slit width of 1 nm, only some 2×10^{-3} of the available light is being used at any one time. Suppose now instead of one spectrometer, we had five hundred acting in parallel, each with a slit width of 1 nm with each monochrometer having a unique wavelength setting. The recording of such a spectrum would then take five hundred times shorter for the same signal to noise. Such a system would be prohibitively expensive, but the powerful arguments in its favor have led to the development of one-dimensional detector arrays to allow a method for reading out the intensity of the different wavelength components (parallel detection). These arrays consist of N photodetecting

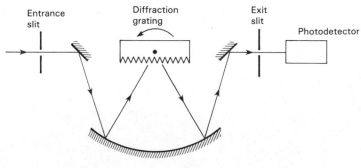

Fig. 3.1

elements in a linear configuration, reducing the scan time by a factor of N (Fig. 3.2). For experimental systems where spatial information is required, as for example in a fluorescence experiment or laser beam profiling, two-dimensional arrays have become available (Compton and Landon 1984).

The photographic plate was one of the earliest means of obtaining both parallel detection and imaging but suffered from the drawback that the plate needed to be developed chemically and did not allow real-time data processing. The photographic plate has been replaced by other devices, and

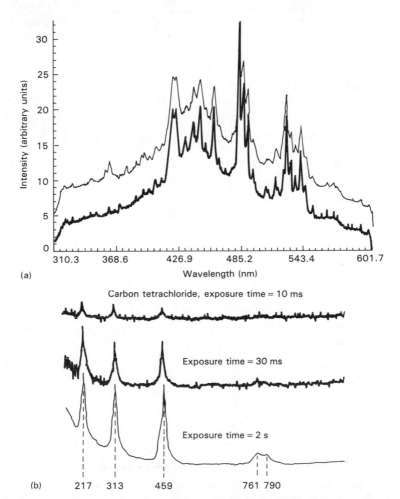

Fig. 3.2 (a) Two spectra taken with a gateable intensified rapid-scan spectral analyzer (70 ns gate width). The dark trace shows the first spectrum, taken 2.09 μs after the flashlamp was triggered. The lighter trace was taken about 250 ns later. Note the change in intensity and spectral content between the two traces; (b) Raman spectra of carbon tetrachloride, obtained by means of signal averaging with a silicon photodiode array. The top trace results from one scan and an exposure time of 10 ms; the middle, from 3 scans totaling 30 ms; the bottom for 200 scans totaling 2 seconds. (Compton and Landon 1984.)

this chapter concentrates on three basic areas: the vidicon; the photodiode array; and the photo MIS (metal insulator semiconductor) capacitor, which includes the charge coupled device (CCD) and the charge injection device (CID).

3.1 THE VIDICON

The vidicon uses photoconductivity as its means for producing an image of the light incident upon it. The photoconductive 'target' is contained in a tube together with a heated cathode or gun which can direct a beam of low velocity electrons onto it. Various coils external to the tube can focus and deflect the electron beam within the tube (Fig. 3.3).

The photoconductive film has a transparent conductive film on it, as shown, which is held at some potential above ground. By means of the deflection coils, the electron beam can be continuously scanned across the surface of the target in much the same way as in television. In the dark, the scanning electron beam can keep its side of the target essentially at the potential of the electron gun, while that side of the target which will eventually be exposed to the light will charge up to the bias voltage because the target is in the high resistance state. We can think of the target as composed of a number of picture elements, or 'pixels', each pixel being effectively a capacitor shunted by a light-sensitive resistor. If a pixel is exposed to light, the resistivity of the element drops markedly, and the charge deposited on it can leak away, i.e. the pixel capacitance is discharged. When the scanning electron beam reaches the discharged pixel, it will recharge its side and a pulse of charge will be obtained from the bias supply which can be read out (the video signal). If the RC time constant of the pixel is very much larger than the time which elapses between two successive scans of the electron beam (known as the frame time, e.g. 1/25 s for a television-type format),

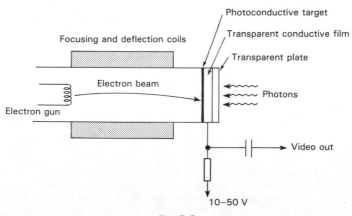

Fig. 3.3

then the video signal will be linearly related to the photocurrent. The light flux is integrated over a frame time which improves the signal-to-noise ratio.

The two-dimensional imaging technique is reasonably sensitive and is near real time. It does however have relatively high dark current, which will limit its use in situations of low light intensity. It also suffers from a phenomenon known as lag. This is the persistence of information on the target between successive scans of the reading electron beam so that some of the optical signal from one scan can be carried over as interference into successive scans. Another problem is blooming, which is the spreading of a strong optical signal from the illuminated area to neighboring areas. This can result in strong signals masking out much weaker signals.

The signal-to-noise problem can be solved with a device known as the plumbicon, which replaces the simple photoconductive target with a thin film p-i-n structure. However, modern vidicon systems use silicon diodes arranged in matrix form as a target. Typically, the diodes will have an active area of ~10 μm with the center to center separation of 25 μm (Fig. 3.4). These devices have extremely short rise times (<5 ps) and largely overcome the problems of lag and bloom, and have a large dynamic range (Cromwell and Labuda 1969),

The light produces electron–hole pairs in the depletion region of the silicon p-n junction, the holes drift towards the surface exposed to the electron beam, and so discharge the photodiodes exposed to the light. The affected diodes are recharged by the scanning beam and the necessary charge will be detected as a video signal as before.

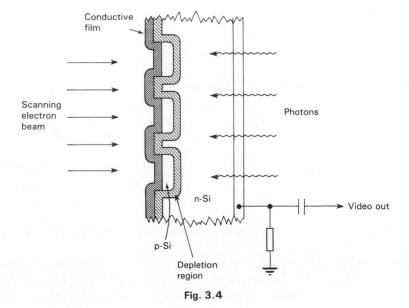

Fig. 3.4

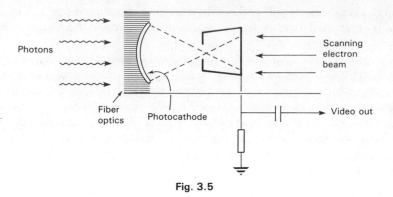

Fig. 3.5

The build-up charge on the silicon dioxide between the diodes prevents complete recharging of the diodes and a carefully controlled leakage path over the oxide has to be used, such as a thin film of antimony trisulphide (SbS_3).

The performance of these silicon vidicons can be improved by cooling or by coupling them to a photocathode to produce the silicon intensified target (SIT) vidicon. In this latter arrangement, photons are projected through a fiber optic array onto a photocathode. An electrostatic lens accelerates the photoelectrons thus produced and focuses them onto the silicon target. The electrostatic lens is gateable (Fig. 3.5).

The sensitivity of the silicon vidicon can be further increased by coupling to an extra intensifier stage such as a microchannel plate (intensified silicon intensified target, or ISIT).

3.2 SOLID STATE SELF-SCANNED ARRAYS
(Weimer 1975, Jespers *et al.* 1976)

In the following sections we will consider all solid state array systems. These systems offer small size, high speed, high reliability and ease of use. Charge carriers are generated by photons in the sensing element and are stored until the sensing element is interrogated. These arrays are categorized on the basis of the means of interrogation. Three basic types can be recognized: the self-scanned silicon photodiode array (SiPDA); the charge coupled device (CCD); and the charge injection device (CID).

3.2.1 Silicon photodiode arrays (SiPDA) (Fry 1975, Weckler 1969)

Arrays of discrete silicon photodiodes fabricated on a single silicon chip are available either as linear or two-dimensional arrays. Linear arrays range in size from 64 to 1024 diodes. Typical 2D arrays are 50×50. Such arrays may be operated in two modes – the continuous mode and the charge storage mode. In the continuous mode, the photocurrent due to the incident photon flux is monitored. This current is very small for most common incident light levels, necessitating sensitive detectors with extremely large gain–bandwidth products in order to read out high definition arrays continuously.

This mode of operation has been superseded by the charge storage mode of operation. In this technique, a photodiode is charged to a reference voltage through some switch and the switch then turned off. From this time until the next time the switch is closed (the integration time), the charge stored in the diode will decay as a result of (a) the diode leakage current, and (b) photocurrent caused by incident illumination during the integration time. At all except very low light levels, decay due to leakage is insignificant and the decay is proportional to the incident light level. At the end of an integration time, the diode capacitance is recharged to its original level and the charge pulse needed for this is equal to the photocurrent integrated over a whole period. It is only necessary to charge the individual diodes in the array sequentially from a common output line so that the charge pulses flowing in this line represents a serial multiplex of the individual integrated photosignals.

Assuming Poisson statistics for the photon emission, the r.m.s. fluctuation in the charge lost by a diode in an integration time is

$$\sqrt{\langle \Delta Q^2 \rangle} = e\eta\sqrt{\langle n \rangle \tau}$$

if there is a mean photon flux of $\langle n \rangle$.

The signal-to-noise ratio is

$$\frac{S}{N} = \sqrt{\langle n \rangle \tau}$$

so that the signal-to-noise ratio increases as the square root of the integration time.

There is, however, a practical maximum to the integration time which is dictated by the diode leakage current under dark conditions. In the dark, the diode will hold its charge sensibly constant for times of the order of milliseconds, depending on the temperature, before leakage causes it to fall and this will then set the maximum integration time, consequently determining the minimum light level which can be detected by the diode array. At high light levels, we must ensure that the diode is not completely discharged during an integration period or else saturation of the diode array will occur.

The sequential readout is performed by MOS transistors embedded in the same monolithic structure containing the photodiode array (Fig. 3.6).

In operation, a voltage pattern is output from the shift register such as to switch on only one transistor at a time. For example, transistor one is first switched on, allowing its associated photodiode to be charged. Next, transistor one is first switched on, allowing its associated photodiode to be charged. Next, transistor one is switched off, and transistor two is switched

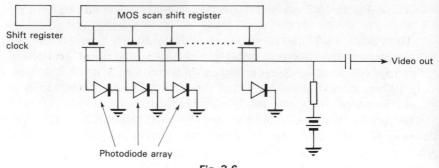

Fig. 3.6

on. The switching voltage is shifted sequentially around all the diodes. The scheme can be extended to two dimensions, as shown in Fig. 3.7.

One row is switched on and all columns then scanned sequentially. The next row is switched on and the columns again scanned sequentially, and so on until all photodiodes have been interrogated.

SiPDAs are chiefly available as one-dimensional arrays. A typical diode linear array will consist of 512 or 1024 photodiodes on 25 μm centers which can be cooled to improve performance.

For small area arraying applications, where resolution need not be of television quality, two-dimensional arrays of these devices can be employed. Present technology and economics do not allow a very high density of imaging cells in SiPDAs and there is now a move to another type of monolithic imaging array to produce full video quality imaging – the charge coupled device.

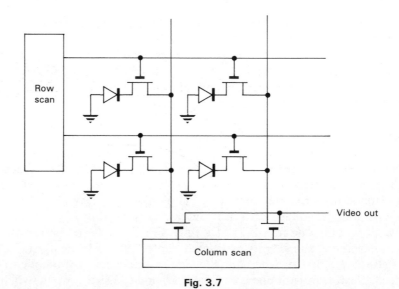

Fig. 3.7

3.2.2 Photo MIS capacitor devices (Boyle and Smith 1971)

The photo MIS (metal insulator semiconductor) device consists of a transparent gate electrode (made of polysilicon or metal oxide), an insulating layer, and semiconductor, as shown in Fig. 3.8.

When a voltage is applied to the gate electrode, with respect to the semiconductor substrate, the energy bands in the semiconductor bend. Let us suppose we have a p-type substrate, and the gate electrode is made positive with respect to the substrate. The holes beneath the electrode are repelled and the energy band diagram is as shown in Fig. 3.9.

The electron potential at the semiconductor–insulator interface under the electrode is lower than in the bulk of the semiconductor by an amount ϕ_{so}, i.e. a potential well of depth ϕ_{so} has been formed at the surface of the semiconductor–insulator interface. Absorption of photons produce electron–hole pairs; the electrons (the minority carriers) will be stored in the potential well. If the contribution of thermally generated electrons can be kept small compared with the number of photogenerated carriers during the light integration period, the charge stored in the device will be directly related to the incident integrated photon flux.

The readout of an array of photo MIS capacitors can be performed in several ways. One way is to transfer the stored charges parallel to the surface of the array to a charge detection system, this mode forming the basis of the charge coupled device. Another way is to inject the stored charge carriers into the bulk semiconductor, producing the charge injection device.

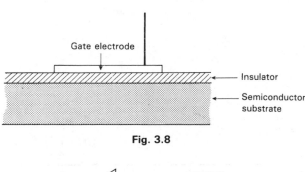

Fig. 3.8

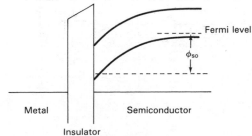

Fig. 3.9

The charge coupled device (CCD) (Barbe 1975)

In some senses, the CCD is similar to the SiPDA. It is fabricated on a single integrated circuit, and any image produced on its surface is scanned off in a serial fashion. CCDs have a smaller cell size than SiPDAs and are more likely to achieve TV quality in a two-dimensional image sensor. More importantly, CCDs have much wider dynamic range than SiPDAs, and have the potential for supplying full video quality imaging at both high and low light levels.

Consider now a p-type substrate of silicon with an array of metal electrodes deposited on the surface, as shown in Fig. 3.10. The electrodes are connected in groups of three and operated with a three-phase voltage supply to give direction to the transfer of accumulated charge. Figure 3.10 shows the way in which charge can be transferred. In (a), V_2 is greater than V_1 so

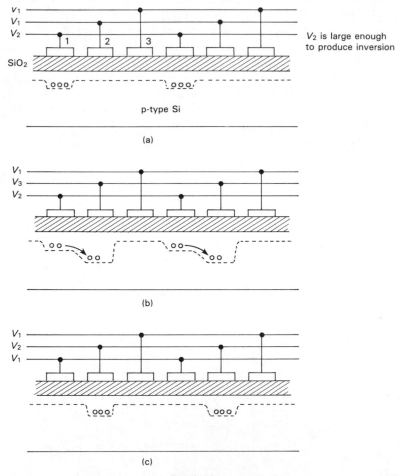

Fig. 3.10

that the potential well beneath the electrode connected to this line is deeper than under electrodes 2 and 3. The photogenerated minority carriers will collect under these electrodes. Now, electrode 2 is made more positive than electrode 1, leaving electrode 3 in its initial condition. Figure 3.10 (b) shows the potential wells beneath the electrodes and the electrodes can now 'funnel' from one well to the next. Finally, we can leave the charge under electrode 2 as shown in (c). By continued application of these voltages, we can move the photogenerated minority charge through the structure.

Such a three-phase clock is not the only scheme that can be employed (Boyle and Smith 1971). A two-phase scheme is illustrated in Fig. 3.11. The problem for clocking schemes with less than three phases is the need to produce a directionality in the structure. This can be achieved, for example, by producing an ion-implanted layer under alternate electrodes as shown. This layer forms a transfer barrier. On the application of a voltage to two electrodes, the potential barrier under the electrode containing the layer is

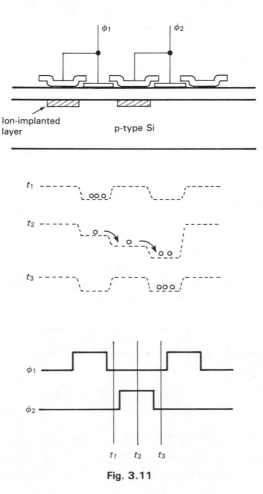

Fig. 3.11

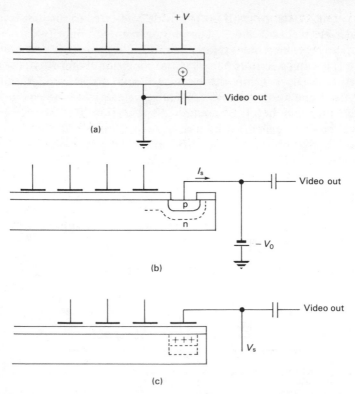

Fig. 3.12 (a) Injection into substrate; (b) p–n reverse-biased junction; (c) floating gate

smaller than that under the other electrode. In this fashion, a direction can be added to the structure as illustrated in the figure.

Several schemes are available to read out the charge stored into the video system. These are illustrated in Fig. 3.12. Figure 3.12 (a) uses a positive voltage on the final electrode to inject holes into the substrate, which is connected to ground via a resistor. When a charge packet is transferred to the last electrode, the hole injection causes a current to flow in the resistor and hence generates a video signal. Figure 3.12(b) uses a reverse-biased diode at the end of the line. The bias voltage $-V_0$ is more negative than any of the surface potentials used for transfer. When charge is transferred to the diode, a current I_s is produced in the output circuit as shown. The final scheme relies on the fact that the capacitance of the MOS structure depends on the charge stored in it. The voltage V_s will therefore depend on the charge under the final electrode (capacitive division).

Organization of CCD image sensors

With a CCD light sensor as previously described, there will be problems if the light is continuously incident on the device. During the readout process,

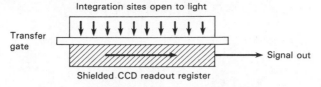

Fig. 3.13

minority carriers are still being generated and information will be super-imposed on the original charge distribution, acting to 'smear' the original image. For small devices having a readout time short compared with the integration time looking at a static image, this is not a serious problem, but otherwise some additional precautions must be taken.

An obvious solution is to protect the device in some manner from the light during readout time. This results in a loss of effective integration time. A more practical solution is to have separate integration and readout sites, as shown in Fig. 3.13.

The CCD photodetectors are located at the side of a CCD readout register, which is shielded from light, and these are separated by a transfer gate. After suitable integration of the light, the transfer gate is opened and charge transferred to the readout register, where it can be clocked out. Except for the short time when the transfer gate is opened, a continuous readout of video information can be obtained.

This philosophy can be used for two-dimensional arrays. Figure 3.14 shows how several linear arrangements can be assembled to form an area array. The vertical scan generator allows the clock pulses to be switched to any linear array which can be read out via the shielded readout register. A better approach is the frame transfer device shown in Fig. 3.15.

In this technique, the integrated signal is passed down into the storage area in the time interval between frame displays (as, for example, during the normal vertical retrace period of a television display system). The whole pattern of charge can be moved down one line at a time into the readout register. If a television-type display is used, one line is moved down during the horizontal retrace period.

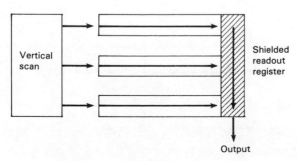

Fig. 3.14

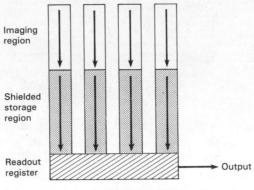

Fig. 3.15

Charge injection devices (CID) (Micken and Burke 1973)

The operation of this device is similar to that of CCDs in that surface potential wells are created by means of external voltages applied to an array of gate electrodes. When photons are absorbed near these potential wells, minority carriers are collected in the wells. After some collection time, the potential well is collapsed and the collected minority carriers are injected in the bulk to combine with majority carriers. This recombination produces a substrate current which is proportional to the collected photoinduced charge. After recombination, the wells are reformed and the collection

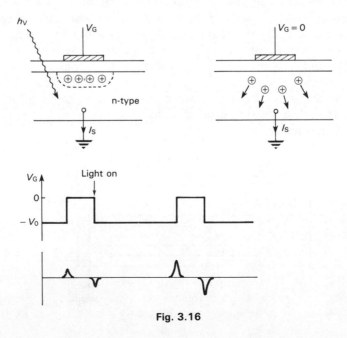

Fig. 3.16

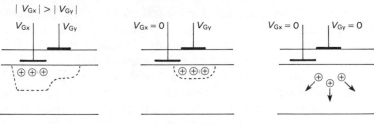

$$| V_{Gx} | > | V_{Gy} |$$

Fig. 3.17

process starts all over again. This process is shown schematically in Fig. 3.16. The collected charge may be sensed by integrating the substrate current during injection into the substrate or by monitoring the voltage on the electrodes.

Two-dimensional CID arrays can be constructed by using two MIS capacitors at each sensing site. Stored charge can be transferred from one capacitor to the other, thus providing a two-axis selection method for scanning. Figure 3.17 shows such a system in operation. Injection of stored charge in a particular sensor will occur only when both electrodes are switched off, thus allowing sampling. It is then possible with horizontal and vertical scan sensors to address a particular element in a random fashion.

3.3 NOISE IN SOLID STATE ARRAYS

There are several sources of noise in solid state arrays. We have dealt in some detail with photon noise from the source and background radiation and shall not deal with these further. In this section we shall look at noise due to scanning of the sensors, and thermal noise associated with the detection amplifiers and reset resistor (Melen 1973).

In SiPDAs and CIDs, we have encountered the use of MOS transistors as analog switches which address the individual photoelements in the array. Due to stray capacitances between the MOS transistors, the photoelements, and the video output line, voltage spikes are obtained on the analog photosignal line being switched whose magnitudes vary throughout the array due to mismatches in the capacitances. These spikes give rise to fixed pattern noise, which occurs in the video passband and so cannot be filtered. The variation of this noise is, however, small compared with the absolute magnitude of the spikes. CCDs are affected by fixed pattern noise resulting from capacitive coupling between clock lines and the output lines. These spikes are all of the same height and can be filtered out. Such filters consume power, and a better method of reducing this parasitically coupled noise is to fabricate video amplifiers directly on the CCD chip. This gives significantly lower parasitic capacitance than CCDs with off-chip amplifiers. Thermal noise in the photoelements can also give rise to fixed pattern noise. We have discussed the physical origins of this noise, and, in arrays, this form of noise is a major problem at light levels below $10 \, \mu W \, m^{-2}$ and for light integration periods of longer than 100 ms. CCDs are more susceptible to this form of noise than SiPDAs.

In CCDs, the main form of noise is known as transfer-loss noise. If enough time is allowed for a transfer of charge, all the charge from one well should be transferred to the adjacent well. In practice, a small fraction ε of the charge in a well is left behind in a transfer. This type of noise appears in the sensed image as a white smear to one side of a sensed white spot, with the biggest smear coming from spots starting furthest away from the output. It is most noticeable when large quantities of charge are being transferred, corresponding to a high-intensity spot. An isolated packet of charge which has undergone n transfers will be reduced by a factor

$$\frac{Q_n}{Q_0} = (1 - \varepsilon)^n \approx 1 - n\varepsilon$$

if the inefficiency product $n\varepsilon$ is very small compared with unity. For example, a two-phase 512-element array with a loss of 10^{-5} per transfer will suffer a reduction in a charge packet of about 1 percent.

The fraction ε also shows statistical fluctuations introducing additional noise. If N_s is the total number of carriers in a signal packet, then, on average, εN_s will be left behind at each transfer. There will be fluctuations about this mean with mean square value $2\varepsilon N_s$, i.e. shot noise is introduced into the signal packet once upon entering and a second time upon leaving a well. If ε is independent of the amount of charge being transferred, the fluctuations at each transfer will be independent, so that the mean square fluctuations will add, i.e.

$$\langle \Delta n^2 \rangle = 2\varepsilon N(N_s + N_b)$$

where N = the number of gates or transfers
N_s = the number of signal carriers per charge packet
N_b = the number of background carriers per charge packet.

At low clocking frequencies ($\ll 1$ MHz), the main cause of transfer inefficiency is traps occurring in the insulator–semiconductor interface. The effects of these trapping interface states are reduced by burying the transfer channels about $1 \, \mu m$ beneath the surface of the substrate (the buried channel device). Bulk traps will play the same role in these devices as surface states, but it is easier to control the density of bulk traps than surface states.

At clock frequencies greater than about 1 MHz, even in the absence of traps, transfer inefficiency can be caused by lack of time for charge to transfer between wells under the influence of self-induced drift, thermal diffusion, and fringing fields (Carnes and Kosonocky 1972).

The final source of noise we shall consider with arrays is due to thermal noise in the amplifiers and the reset resistor. The noise at the output of the amplifiers is a function of the source impedance of the amplifier and the

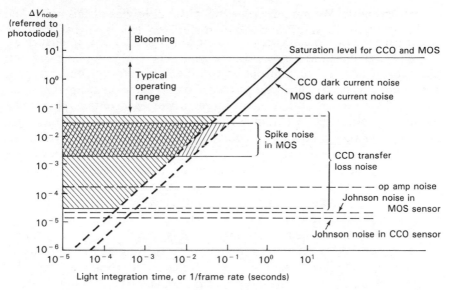

Fig. 3.18 (From Melen 1973)

noise parameters of that amplifier (see Chapter 4 on amplifiers and filters). Since the source (the photosensitive element) is effectively a capacitance, the larger the capacitance the greater the noise. Since the SiPDA and the CID array have their outputs connected to high capacitance bus lines, this noise is greater in these devices than in CCDs. In fact, the favorable noise characteristics of the CCD array arise from the fact that the capacitance of each stage of the registers leading from the sensor element to the output amplifier on the same chip is no greater than the capacitance of the photoelement itself.

Reset noise represents the effect of recharging the capacitance C of the photoelement through a noisy resistor R at the end of an integration period. It can be shown (Carnes *et al.* 1973) that the mean square charge on the capacitor after recharging is given by

$$\langle \Delta Q^2 \rangle = kTC$$

and this is the noise charge superimposed on the video signal due to this effect.

The various noise sources in solid state arrays and their relative importance as a function of integration time are compared in Fig. 3.18, taken from Melen (1973).

3.4 IMAGING IN THE INFRARED (Tebo 1984)

Vidicon tubes have been constructed out of photoconductive materials which are responsive in the infrared out to 3 μm. Beyond this wavelength, photoconductive materials are too conductive for use in vidicons. To over-come this problem, pyroelectric vidicons have been produced.

The pyroelectric vidicon uses a thin slice of pyroelectric material, such as triglycerine sulphate, as a target. The target is, however, insulating and not photoconductive; its mode of operation differs, therefore, from that of a conventional vidicon. The image produced on the target of the pyroelectric vidicon induces a temperature on the target, thereby generating a charge via the pyroelectric effect. A scanning electron beam discharges the target to cathode potential, producing a video signal as in conventional vidicons. However, because the target is not photoconducting, the surface of the target facing the cathode is rapidly reduced to cathode potential, and, in the absence of any change in the image, the pyroelectric vidicon will produce no signal. To produce a continuous signal from such a vidicon, two methods are used. One uses an optical arrangement so that the image is con-tinually scanned over the target, while ensuring that the electronic scan of the target follows this motion to produce a continuous stationary scene at the display. The other method involves mechanically 'chopping' the light from the image falling on the target so that, on alternate scans, a signal is generated proportional to the difference between the radiation from the scene and the chopper.

Silicon is the obvious material for both photodiode arrays and photo-MIS devices since it allows integration of photosensor and scanning devices and because of the highly developed MOS device technology. However, intrinsic silicon responds only to wavelengths less than 1.1 μm, while for wavelengths less than 0.3 μm the photoresponse falls off due to the high ab-sorption coefficient, which causes carriers to be generated almost entirely in the surface where there are high recombination rates. Several approaches are being investigated to extend the response of arrays into the infrared (SPIE Conference 443, *Infrared Detectors*, San Diego, California, August 1983).

1. *Extrinsic silicon detector arrays.*   Extrinsic silicon devices are capable of detection between 2 and 20 μm. The major difference between these devices and intrinsic devices is the much lower absorption coefficient and the need to operate the extrinsic devices at much lower temperatures than intrinsic devices. In extrinsic silicon operating in the infrared, only majority carriers are generated, so that the depletion mode MIS struc-ture cannot be used directly as a sensor. A solution to this problem is to use a hybrid device, where photons are detected in the extrinsic silicon while charge is stored in an intrinsic MIS device (Barbe 1977).
2. *Schottky barrier CCD arrays.*   Schottky barrier array devices operate in the near and middle infrared. Since it is the only infrared technology compatible with standard integrated circuit processing, these devices can be produced in large volume and at low cost. PdSi Schottky barrier

CCDs have a cut-off wavelength of 3.6 μm with a quantum efficiency of up to 8 percent. PtSi devices operate between 1 and 6 μm and, operating in staring mode, can have integration times of 10 to 100 ms at a temperature of 80 K. High quality imaging has been achieved with a 64×128 PtSi Schottky barrier infrared camera operated at 60 frames per second.

3. *Indium antimonide detector arrays.* Indium antimonide is a reliable and popular detector in the 1 to 5 μm range. There are problems with MOS technology in this material, so that arrays are usually produced using a hybrid technology. The photoelements are constructed out of indium antimonide and these diodes are individually connected to the multiplexor or readout mechanism which is constructed out of a well-developed technology such as silicon. Progress has been made in the development of MOS technology in indium antimonide, using deposited silicon oxynitride as an insulator, and linear all-indium antimonide arrays have been developed. CIDs using indium antimonide, for both linear and area arrays, have been developed. The arrays are operated by a silicon MOS scanner that controls the gate voltages and provides sequential access to the photo-MIS devices.

4. *Cadmium mercury telluride.* Cadmium mercury telluride arrays are employed in the spectral range 2 to 25 μm. Linear arrays using photoconductivity are now in full production. Area arrays are usually fabricated using hybrid technology. The device technology of cadmium mercury telluride has now developed sufficiently to allow some integration of the photosignal to occur in the cadmium mercury telluride itself before transfer to a silicon CCD.

For broadband infrared work pyroelectric arrays are now readily available, and, although currently more expensive than SiPDAs, are less expensive than other semiconductor infrared arrays. They are about 10^5 times less sensitive than SiPDAs but, because of the way pyroelectric devices operate, they are insensitive to ambient light, an advantage over silicon devices. Because of this insensitivity to c.w. light, c.w. sources need to be optically chopped, with a multiplexing rate fast enough to scan all the elements in a time short compared with the chopping cycle time, thus ensuring all the elements have the same integration time. We should note with pyroelectric arrays that the array should be scanned twice during a chopping cycle. As with SiPDAs, when the light is incident on a pyroelectric element, positive charge is induced by the pyroelectric effect. This charge can be fed into an external circuit as a voltage whose magnitude is proportional to the incident light intensity. When the light is blocked (by the chopper) during the second half of the chopping period, the pyroelectric element cools down, creating negative charge, which must be taken away to rezero the device – hence the second scan (Roundy 1983).

3.5 IMAGING MODES

In this section we shall look at some of the techniques used for imaging light with arrays.

3.5.1 Staring mode

In this mode of operation, the imaging array integrates the field of view for some period of time, after which the field is read out while a second field is being integrated. Let us suppose we have an array of elements of mean responsivity R and let the deviation of the response of the ith element from the mean value be ΔR_i. Further suppose that the average number of photons incident on the array is N and that the contrast of the image is C. After integrating for τ seconds, the number of electrons in the ith element of the array is

$$N_i = (R + \Delta R_i)N(1 + C)\tau$$

$$= NR\tau + NCR\tau + N \Delta R_i\tau + N \Delta R_i C\tau$$

We can ignore the final term as negligible. The first term is a constant giving the average number of photogenerated carriers in each element of the array. The second term is the desired signal term, while the third term is a noise term, the element-to-element variation in the number of carriers due to spatial inhomogeneities. If the standard deviation in this noise term is greater than the signal term, then the image will not be detectable. However, this nonuniformity in response is a fixed function and so can be measured and stored in the memory of a computer. This information can then be used to subtract off the element-by-element nonuniformity to improve the quality of the image.

3.5.2 Parallel scanned

A second mode of imaging is known as parallel scanning, in which a linear array of sensor elements is used together with rotating optics to provide parallel scanning of the scene across the array. This is illustrated in Fig. 3.19.

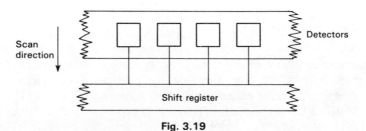

Fig. 3.19

The outputs of the dectector array may be multiplexed; for example, with CCDs. Once per horizontal line time, the CCD can sample the detector output voltages and shift out the resulting signal.

3.5.3 Serial scanned

In the serial scanned mode, one sensor element is used to raster scan an image field. This provides a better uniformity than the parallel scanned mode but is not as sensitive. CCD delay-and-add techniques were subsequently applied to the serial scanned mode and this is known as *time delay and integration* (TDI). TDI techniques are now usually implied in the term serial scan.

The TDI technique can be illustrated with reference to Fig. 3.20. In this diagram we have a CCD consisting of M columns each containing N elements. The array has a velocity, relative to the object to be imaged, in the direction parallel to the individual columns of the array.

If the charge is clocked out of the array at such a rate that the transfer of a charge packet along the CCD columns is synchronous with the corresponding point in the image plane of the object, then the charge generated by illumination from the same point on the image will add together in successive elements along a CCD column. Thus, after N elements in the column, the signal from some point on the object will be N times that due to a single element without the TDI technique. However, the noise accumulated during the transfer will add incoherently, and after N bits is $\sqrt{N}$ times the single element noise without TDI. Thus, TDI will bring about an enhancement of the signal to noise given by $\sqrt{N}$ over the single detector case.

It is possible to achieve a mixture of serial and parallel scanning techniques. Such a scheme involves parallel optical scanning and serial TDI. This scheme requires a two-dimensional detector array, with the number of elements perpendicular to the scan direction determining the resolution of the device.

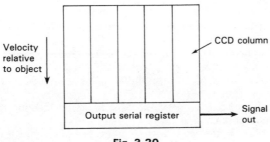

Fig. 3.20

SUMMARY

In this chapter we have looked at some of the devices which can be used for the imaging of light or for the simultaneous recording of the intensity of the different wavelength components of an optical spectrum. The chapter has concentrated on solid state devices, such as the MOS photodiode array or the charge coupled device (CCD), which are small and fast, operate at low powers and low voltages, and have high dynamic ranges. Such devices are available as linear arrays or area arrays. Both types of array are similar in operation, with similar fabrication methods. Photons generate charge carriers in the sensing elements where they are stored and subsequently interrogated. We have categorized the arrays on the basis of the means of interrogation. In the photodiode array, the charge is collected by the junction of a reverse-biased photodiode and the stored charge is read out by a single charge transfer from the junction to the external video circuitry. In the CCD and the CID (charge injection device), the photogenerated charge is collected and stored in an electric field induced potential well beneath an electrode. In the case of the CCD, readout is accomplished by means of multiple transfers of charge through the array of induced potential wells to the output circuitry. For the CID, the wells are collapsed and the resultant substrate current is sensed.

REFERENCES

Barbe, D.F. 1975. Imaging devices using the charge coupled concept . *Proceedings of the IEEE,* Vol. 63, pp. 38–67.

Boksenberg, A. 1982. Advances in detectors for astronomical spectroscopy. *Philosophical Transactions of the Royal Society*, London, Vol. A 307, pp. 531–42.

Boyle, W.S. and Smith, G.E. 1971. Charge coupled devices – a new approach to MIS device structures. *IEEE Spectrum,* July, pp. 18–27.

Carnes, J.E. and Kosonocky, W.F. 1972. Noise sources in charge coupled devices. *R.C.A. Review,* Vol. 33, pp. 327–43.

Compton, R.D. and Landon, D.O. 1984. The role of parallel detection in spectroscopy. *Lasers and Applications*, August, pp. 65–8.

Cromwell, M.H. and Labuda, E.F. 1969. The silicon diode array camera tube. *Bell System Technical Journal*, Vol. 48, pp. 1481–528.

Fry, P.W. 1975. Silicon photodiode arrays. *Journal of Physics E: Scientific Instruments*, Vol. 8, pp. 337–49.

Jespers, P. G., van der Wiele, F. and White, M. H. 1976. Solid state imaging. Nato Advanced Study Institute Series, Noordhoff-Leyden.

Melen, R. 1973. The trade-off in monolithic image sensors: MOS vs CCD. *Electronics*, Vol. 46, pp. 106–11.

Michen, G.J. and Burke, H.K. 1973. Charge injection imaging. I.S.S.C.C. Dig. Tech. Papers, February.

Roundy, C.B., 1983. Pyroelectric arrays make beam imaging easy. *Lasers and Applications*, January, pp. 55–60.

Tebo, A. 1984. I.R. detector technology part II: arrays. *Laser Focus,* July, pp. 68–82.

Weckler, G.P. 1967. Operation of p-n junction photodetectors in a photon flux integration mode. *IEEE Journal of Solid State Circuits,* Vol. 2, pp. 65–73.

Weimer, P.K. 1975. Image sensors for solid state cameras. *Advances in Electronics and Electron Physics,* Academic Press, Vol. 37, p. 197.

PROBLEMS

3.1 An image detector is used to view the earth in order to be able to resolve objects having a temperature of 1 K above the average background temperature of 300 K. The detector material has a cut-off wavelength λ_c. Assuming that the earth behaves as a perfect black body, calculate the contrast the detector must have to resolve the objects when:

(a) $\lambda_c = 2 \ \mu m$
(b) $\lambda_c = 10 \ \mu m$.

Note: The contrast is given by

$$C = \int_0^{\lambda_c} \frac{1}{Q} \frac{dQ}{dt} \, d\lambda$$

where Q is the spectral photon flux

$$Q = \frac{2\pi c}{\lambda^4} \frac{1}{\exp(hc/\lambda kT) - 1}$$

3.2 An infrared imaging system is used in the staring mode and must distinguish objects having a temperature difference of 0.2 K in a spectral region where the contrast is 5% K^{-1}. Calculate the maximum allowable deviation of the response of the elements in the array from the mean.

4

Amplifiers and Filters

We have seen how we may transform optical information into an electrical signal. The signal information needs to be preprocessed in order that it may be in the correct form for analysis. Probably the most important element in the preprocessing system is a signal amplifier. Although this amplifier may be constructed out of discrete active components such as transistors and FETs, the most popular device used for this purpose is the operational amplifier, an all-solid-state integrated circuit. It is this device which we shall study in this chapter.

The reason for the popularity of this device is its ability to amplify d.c. signals and a.c. signals simultaneously without phase shift due to interstage coupling. It is a direct-coupled device with inverting (−) and noninverting (+) inputs and a single-ended output. It has a common earth terminal and operates with a bipolar power supply. It has a high (usually greater than 10^5) voltage gain, a high input resistance, and a low output impedance.

4.1 BASIC OPERATIONAL AMPLIFIER CONFIGURATIONS (Schick 1971)

We can distinguish two different configurations of amplifier. The first is the inverting configuration, where a signal at the − input produces a negative-going signal at the output. The second is the noninverting configuration, where a signal at the + input produces a positive-going signal at the output. These are shown in Fig. 4.1.

The expressions for the overall gain of the amplifier system have been obtained for the so-called ideal operational amplifier having infinite gain and infinite input impedance. In this infinite gain approximation (IGA), no current flows into the inverting or noninverting terminals of the op-amp and the voltage difference between these terminals is zero. This simplifies greatly the analysis of op-amp circuits and can be illustrated by a consideration of the differential amplifier shown in Fig. 4.2. This circuit has signal applied to both terminals and is a combination of the inverting and noninverting amplifier.

Since no current flows into the + terminal, it follows directly that

$$E^+ = \left(\frac{R_{f_2}}{R_{f_2} + R_2}\right) \cdot E_2$$

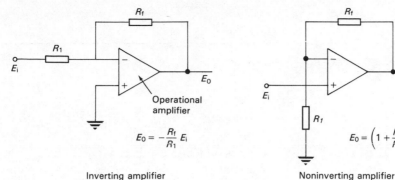

$$E_0 = -\frac{R_f}{R_1} E_i$$

Inverting amplifier

$$E_0 = \left(1 + \frac{R_f}{R_1}\right) E_i$$

Noninverting amplifier

Fig. 4.1

and since the voltage difference between the $+$ and $-$ terminals is zero

$$E^- = E^+ = \left(\frac{R_{f_2}}{R_{f_2} + R_2}\right) \cdot E_2$$

As no current flows into the $-$ terminal

$$i_1 = i_f$$

or

$$\frac{(E_1 - E^-)}{R_1} = \frac{(E^- E_0)}{R_{f_1}}$$

so that

$$E_0 = \left(\frac{R_1 + R_{f_1}}{R_1}\right) \cdot \left(\frac{R_{f_2}}{R_2 + R_{f_2}}\right) \cdot E_2 - \frac{R_{f_1}}{R_1} E_1$$

If now $R_{f_1} = R_{f_2} = R_f$ and $R_1 = R_2 = R$, then

$$E_0 = \frac{R_f}{R} (E_2 - E_1)$$

so that the output voltage is a gain factor multiplied by the differential input

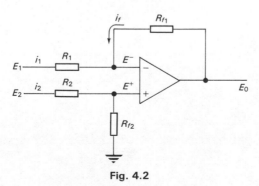

Fig. 4.2

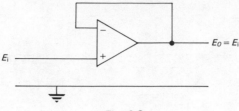

Fig. 4.3

voltage. This configuration is important because it can reject, to a great extent, a signal common to both inputs. This is known as common mode rejection and will be dealt with in more detail later. Such common mode rejection is particularly important where transducers are located some distance from the amplifier. In this case, induced pickup in leads (especially 50 Hz pickup) and ground loops can give rise to signals many times larger than the transducer signal which is being measured. This unwanted component is transformed into a common mode signal, i.e. a signal common to the inverting and noninverting input, which can then be rejected.

Another important amplifier configuration is derived from the noninverting system and is shown in Fig. 4.3.

This is known as a unity gain buffer, and the output voltage follows the input voltage (hence the name 'voltage follower' sometimes given to this circuit). The expression for the output voltage follows directly from the approximations used in the infinite gain approximation. The input impedance of the circuit is effectively that of the op-amp (typically 1 MΩ or greater) while its output impedance is very low (typically about 1000 Ω). Such a circuit is useful as a buffer between a signal source and an amplifier of low input impedance and helps to prevent excessive loading of the signal.

Before leaving this section, two more amplifier configurations can be usefully introduced. The first is the current-to-voltage converter which is used with photodetectors such as photomultipliers and photodiodes. These photodetectors can be considered as current sources with very large output impedances. This circuit is shown in Fig. 4.4.

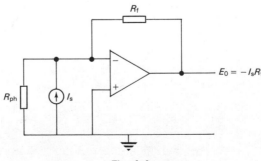

Fig. 4.4

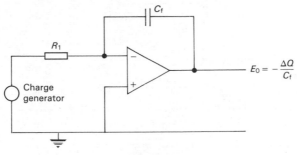

Fig. 4.5

The photodetector is modeled as a current source with its output imped-ance R_{ph}. Because the $+$ and $-$ terminals are at the same potential, no current flows through R_{ph} and we are measuring the short circuit photo-current (the $+$ terminal is at ground). Furthermore, since no current flows into the $-$ terminal, then all the current must flow through R_f, so that the output voltage is

$$E_0 = -I_s R_f$$

For example, if a photodiode produces a current of 10 μA and $R_f = 100$ kΩ, then the output voltage of the current-to-voltage converter would be

$$E_0 = 10 \ \mu\text{A} \times 100 \text{ k}\Omega = 1 \text{ V}$$

The second configuration is the charge amplifier, which would be used, for example, with a piezoelectric transducer. The circuit is shown in Fig. 4.5. Here, the charge ΔQ generated by the transducer flows onto the capacitor C_f, generating a voltage $E_0 = -\Delta Q / C_f$.

The attraction of such circuits for the designer of a signal processing system is that properties such as gain, and input and output impedance, are determined by the external network of resistors and capacitors and are essentially independent of the internal parameters of the operational amplifier. The resistor connected between the output of the op-amp and its inverting input introduces negative feedback into the amplifier. Negative feedback brings many advantages, such as gain stability, but can introduce some disadvantages, such as reduction in the bandwidth of the amplifier. Such drawbacks will be discussed in the section on amplifier errors.

4.2 THE INFINITE GAIN APPROXIMATION (IGA)

In dealing with op-amp circuits, we have been using certain approxima-tions, and in this section the nature of these approximations will be briefly discussed. Let us consider a simple inverting amplifier configuration, as shown in Fig. 4.6, where the op-amp has an input impedance Z_{in} and a gain $-A$.

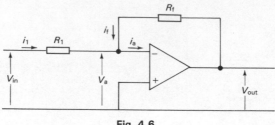

Fig. 4.6

We now cannot ignore the current flowing into the amplifier. Equating currents at the summing junction

$$i_a = i_1 + i_f$$

$$\therefore \quad \frac{v_a}{Z_{in}} = \frac{(v_{in} - v_a)}{R_1} + \frac{(v_{out} - v_a)}{R_f}$$

Remembering that the voltage gain of the op-amp relates the voltage at the input terminal of the op-amp to that at the output

$$v_{out} = -A v_a$$

Eliminating v_a

$$\frac{v_{out}}{v_{in}} = \frac{-1}{\dfrac{1}{A}\left(1 + \dfrac{R_1}{R_f} + \dfrac{R_1}{Z_{in}}\right) + \dfrac{R_1}{R_f}}$$

In the infinite gain approximation, $A \to \infty$ and

$$\frac{v_{out}}{v_{in}} = -\frac{R_f}{R_1}$$

Thus, the infinite gain approximation will be valid if

$$\frac{1}{A}\left(1 + \frac{R_1}{R_f} + \frac{R_1}{Z_{in}}\right) \ll \frac{R_1}{R_f}$$

We can use the above equations to obtain an expression for v_a

$$v_a = \frac{v_{in}}{1 + \dfrac{R_1}{Z_{in}} + A\,\dfrac{R_1}{R_f} + \dfrac{R_1}{R_f}} \approx v_{in}\,\frac{R_f}{AR_1}$$

using the inequality obtained for the validity of the IGA, and so

$$i_a = \frac{v_a}{Z_{in}} = v_{in}\,\frac{R_f}{AZ_{in}R_1}$$

Typically, A will be of the order of 100 000, Z_{in} will vary from $\sim 10\ \text{M}\Omega$ for bipolar input op-amps to $10^{12}\ \Omega$ for FET input op-amps, while R_f/R_1 will be of the order of 10 to 100. Thus the inequality for the IGA is easily satisfied. Furthermore,

$$i_a = \frac{v_{in}}{R_1} \cdot \frac{R_f}{A Z_{in}} = i_{in} \frac{R_f}{A Z_{in}}$$

so that i_a will be negligible compared with i_{in}. Thus we are justified in ignoring the current flow into the terminals of the op-amp.

4.3 AMPLIFIER ERRORS

The discussion so far has assumed an ideal op-amp. For example, if zero input voltage is applied to the inverting amplifier configuration, then zero output voltage should be obtained. Further, it has been assumed that an a.c. signal of any frequency and any amplitude will be amplified without phase shift or distortion. The differences between the ideal and the real op-amp will now be discussed.

4.3.1 D.C. errors

Here, the sources that cause the addition of d.c. error terms to the d.c. output voltage will be illustrated. The characteristics of a real op-amp that introduce d.c. errors are (a) input bias currents, and (b) input offset voltages.

So far, it has been assumed that no current flows into the terminals of the op-amp. However, internally the op-amp has transistors which must be biased correctly for proper operation. For this reason, the + and − terminals of the op-amp will carry small currents, known as the input bias currents. Ideally, these bias currents will be equal for both terminals, but, in practice, there will be a slight difference between them. The difference in magnitude between the bias current in the + terminal I_B^+ and that in the − terminal I_B^- is called the input offset current. Manufacturers' data sheets quote the average input bias current

$$I_B = \tfrac{1}{2}(I_B^+ + I_B^-)$$

and the input offset current

$$I_{os} = |I_B^+| - |I_B^-|$$

for the device. For example, for the popular 741 op-amp, typically $I_B = 80$ nA and $I_{os} = 20$ nA.

Op-amps having FETs in the input stage have the lowest input bias currents, typically picoamps.

To gauge the effect of the bias current on the d.c. output voltage, consider the inverting amplifier having zero input voltage shown in Fig. 4.7(a). The current I_B^- flows through R_f, producing an output voltage

$$E_0 = R_f I_B^-$$

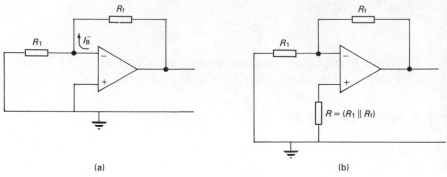

Fig. 4.7

and will be greater as the value of the feedback resistor increases. For example, for a 741 op-amp with $R_f = 1$ MΩ, $E_0 = 70$ mV.

This error can be reduced, as shown in Fig. 4.7(b), by placing a resistor between the + terminal and earth of value equal to the parallel resistance of R_1 and R_f. The voltage at the + terminal is now

$$E^+ = I_B^+ R$$

and, in the infinite gain approximation, this is the voltage at the − terminal. A current $I_B^+ R/R_1$ flows through R_1 to earth so a current

$$I_B^- - I_B^+ \left(\frac{R}{R_1}\right)$$

flows through the feedback resistor R_f. The output error voltage is now

$$E_0 = R_f\left(I_B^- - I_B^+ \frac{R}{R_1}\right) - I_B^+ R$$

$$= R_f\left[I_B^- - I_B^+ R\left(\frac{1}{R_1} + \frac{1}{R_f}\right)\right]$$

If $1/R = 1/R_1 + 1/R_f$, then this reduces to

$$E_0 = R_f(I_B^- - I_B^+)$$

For the example of the 741 quoted above, $E_0 = 20$ mV, which is a significant reduction. Obviously, the smaller the input offset current the smaller the error output voltage. This discussion is easily extended to the other amplifier configurations.

Even if the effects of the bias currents can be ignored (and, in practice, this can be done by reducing the value of the feedback resistor), an output voltage will still be obtained with zero input. The voltage which needs to be applied between the input leads of the amplifier to obtain zero output voltage is known as the input offset voltage.

For the 741 op-amp, typically this value is 1 mV. This can be modeled

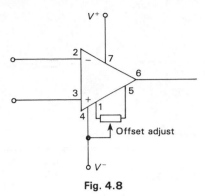

Fig. 4.8

as a battery in series with one of the terminals of the op-amp. This input offset voltage is amplified to produce an error output voltage, e.g. for an amplifier having a gain of 100, an output error voltage of 100 mV is obtained. The input offset voltage may be nulled by introducing an opposite voltage at one of the amplifier inputs or by the manufacturer's internal offset adjustments. In this latter case, usually a simple resistor network connected to some pins of the device will be sufficient, as shown in Fig. 4.8 for the 741 op-amp. This mode will differ between different types of op-amp and the manufacturer's data sheet should be consulted for the preferred method of offset adjustment.

Even if the input offset voltage has been nulled and the effects of the input bias currents minimized, the input offset bias current and voltage will be temperature dependent, i.e. they exhibit drift. The d.c. error output voltage is minimized at one temperature and will increase if the temperature changes. If the change in error output voltage over the range of ambient temperature is excessive, then either the temperature must be controlled more closely or a different op-amp with smaller drift chosen. In addition, drift will occur with fluctuations in the power supply; these can be eliminated by the design of a well-regulated power supply.

4.3.2 A.C. errors

In a discussion of a.c. errors, it is necessary to distinguish between small (less than about a volt) a.c. output voltages and large (greater than about a volt) a.c. output voltages.

For small output, a.c. errors are introduced because the operational amplifier exhibits a voltage gain which varies with frequency. Above a certain frequency, the voltage gain falls (or 'rolls off') due to internal capacitance of the op-amp. Above this frequency, the phase of the output voltage with respect to the input voltage shows a change. The variation of the op-amp gain with frequency is shown in Fig. 4.9. The curve labeled A_{OL}

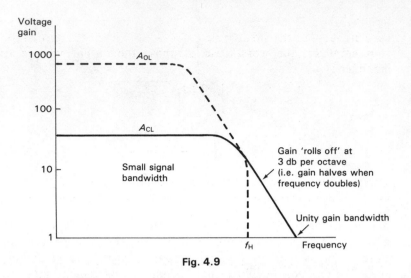

Fig. 4.9

is the curve normally quoted in the manufacturer's data sheet. It shows the *open-loop gain* of the amplifier, i.e. the gain with the negative feedback loop open-circuited. Marked on the curve is the frequency at which the open-loop gain falls to unity. This often-quoted figure is known as the *small signal unity gain bandwidth*. For the 741 op-amp, the unity gain bandwidth is 1 MHz. In actual circuits, the designer is interested in the closed-loop gain of the amplifier, A_{CL}. Again, the voltage falls off above a certain frequency. If the frequency at which the closed-loop voltage gain falls to $1/\sqrt{2}$ of its value at lower frequencies where its response is flat (the midband gain) is f_H, then the small signal bandwidth of the op-amp is defined as

$$\Delta f = f_H - 0$$

It is related to the open-loop unity gain bandwidth by

$$\Delta f = \frac{\text{Unity gain bandwidth}}{\text{Closed-loop gain}}$$

Thus for a 741 with a closed-loop gain of 100 the bandwidth is

$$\Delta f = (1\ \text{MHz}/100) = 10\ \text{kHz}$$

If this amplifier were to be used above this frequency of 10 kHz, errors would be introduced into the output signal.

The above discussion is an illustration of the general principal that the product of the closed-loop gain and the small signal bandwidth is constant. We have therefore to trade off between high gain and high bandwidth. If high gain and high bandwidth are desired, then several low gain, high bandwidth amplifiers should be cascaded.

Even at frequencies below f_H, errors can be introduced into the output voltage if that output tries to change too quickly. This is called slew-rate limiting and is due to limitations in the op-amp internal circuitry to drive capacitance. For the 741, the slew rate is 0.5 V/μs and is quoted for a unity gain amplifier, the worst case. Suppose the output voltage from the amplifier is

$$E_0 = E_m \sin \omega t$$

so that

$$\frac{dE_0}{dt} = \omega E_m \sin \omega t$$

The maximum rate of change of the output voltage is ωE_m. Thus, a given sinusoidal input voltage will be followed faithfully by the output in an op-amp based amplifier if

$$\omega E_m \leqslant \text{slew rate}$$

Consider a 741-based amplifier having a gain of 100 and an input sinusoidal voltage of amplitude 100 mV. The small signal bandwidth is 10 kHz. This implies that a signal frequency of about 10 kHz can be faithfully amplified. However, at this frequency,

$$\omega E_m = 6.28 \times 10^4 \times 10 = 0.628 \text{ V}/\mu\text{s}$$

This is greater than the slew rate, so that for this amplifier it is the slew rate and *not* the small signal bandwidth that limits the high frequency performance of the amplifier.

4.4 THE INSTRUMENTATION AMPLIFIER
(Jaquay 1973, Graeme 1980)

If we wish to measure a low level signal, particularly from remote transducers, then there are several factors we would seek in our amplifier system — sufficient gain, good stability, high input impedance and good common mode rejection.

Let us look more closely at common mode rejection in amplifiers. One way of looking at the amplifier is to consider it as two amplifiers of gain A_1 and A_2, as shown in Fig. 4.10.

The output of this amplifier is

$$V_0 = A_2 V_2 - A_1 V_1$$

For the ideal operational amplifier

$$|A_1| = |A_2|$$

but, in general, this will not be so. We can write

$$V_0 = A_d(V_2 - V_1) + A_{cm}(V_1 + V_2)$$

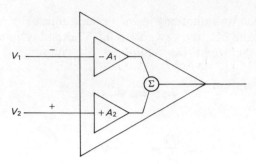

Fig. 4.10

where $A_d = \frac{1}{2}(A_1 + A_2)$

$A_{cm} = \frac{1}{2}(A_2 - A_1)$

A_d is known as the differential gain while A_{cm}, the common mode gain, gives a measure of the departure from ideality and how much of the common mode signal will appear at the output of the amplifier. In general, it is desirable that $A_d \gg A_{cm}$. We can define the common mode rejection ratio as

$$\text{CMRR} = 20 \log_{10} \left| \frac{A_d}{A_{cm}} \right|$$

which is usually a large number exceeding 50 db.

Consider now the differential amplifier shown in Fig. 4.11. In terms of the input voltages V_{i_1} and V_{i_2},

$$V_1 = \left(\frac{1}{R_1} + \frac{1}{R_2} \right)^{-1} \left(\frac{V_{i_1}}{R_1} + \frac{V_0}{R_2} \right)$$

$$V_2 = V_{i_2} \frac{R_4}{R_4 + R_3}$$

assuming the operational amplifier has infinite input impedance. It is then

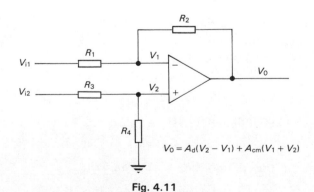

$$V_0 = A_d(V_2 - V_1) + A_{cm}(V_1 + V_2)$$

Fig. 4.11

straightforward to show that

$$V_0 = \frac{R_2}{R_1} \left| V_{i_2} \left[\frac{(A_d/A_{cm}) + 1}{(A_d/A_{cm}) - 1} \right] \left[\frac{1 + (R_1/R_2)}{1 + (R_3/R_4)} \right] - V_{i_1} \right|$$

using the approximation

$$1 \gg \left| \frac{1}{(A_d - A_{cm})} \cdot \left(1 + \frac{R_2}{R_1} \right) \right|$$

These equations illustrate two problems. The ratio

$$\frac{1 + (R_1/R_2)}{1 + (R_3/R_4)}$$

represents a common mode error due to nonmatching of resistors. If the resistors are not perfectly matched such that

$$\frac{R_1}{R_2} = \frac{R_3}{R_4}$$

then a common mode input signal gives rise to a differential signal at the amplifier input terminals and so to an amplified output. Even if this matching occurs, there will be a common mode error due to the amplifier itself. This error is

$$\frac{(A_d/A_{cm}) + 1}{(A_d/A_{cm}) - 1}$$

Suppose we have matched resistors and an operational amplifier with a CMRR of 50 db. Then

$$\frac{A_d}{A_{cm}} \approx 316$$

so that

$$V_0 = \frac{R_2}{R_1} [V_{i_1} - 1.0063 \, V_{i_2}]$$

This is not the only problem with this circuit. If a high gain is to be achieved from the circuit, R_1 and R_3 must be low; low values are undesirable because this will cause excessive loading of the signal. High values cause increased drift and offset. Furthermore, because of the need to match resistors, it is difficult to vary the gain continuously.

We have two routes by which we can proceed. The first is to use more than one operational amplifier to try to overcome the problems mentioned above. A typical circuit is shown in Fig. 4.12, which uses voltage followers with very high input impedance (e.g. with FET input stages).

The second approach is to use an instrumentation amplifier. An instrumentation amplifier is a closed-loop gain block having a differential input, high input impedence, high CMRR and accurately specified gain. In

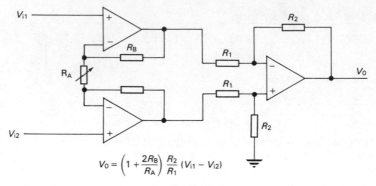

$$V_0 = \left(1 + \frac{2R_B}{R_A}\right) \frac{R_2}{R_1} (V_{i1} - V_{i2})$$

Fig. 4.12

contrast, an operational amplifier is an open-loop gain device used to drive external feedback networks which define the performance of the circuit, and may be single ended or differential. The feedback networks must be supplied by the user in the operational amplifier device, but these are provided in the instrumentation amplifier module itself. The user only adds a gain setting resistor. Instrumentation amplifiers find application in situations where we need to amplify accurately small differential signals having large common mode signals. Typically, they will be found as transducer amplifiers for thermocouples and thermistor networks.

4.5 NOISE IN AMPLIFIERS (Letzer and Webster 1970)

When we feed the signal from the photodetector through the amplifier system, we must take account of the fact that a practical amplifier will degrade the input signal by adding noise to that already present. This is an inevitable consequence of the fact that the amplifier contains active components, like transistors, and passive components, like resistors, which can act as sources of noise. Regardless of whether the amplifier is constructed

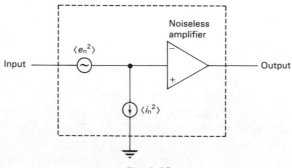

Fig. 4.13

Table 4.1 Typical values for noise sources

Type	$\sqrt{\langle e_n^2 \rangle}$ nV/Hz$^{1/2}$	$\sqrt{\langle i_n^2 \rangle}pA/Hz^{1/2}$
Op-amp 741 (bipolar)	60	4
Op-amp 725 (FET)	9	0.15
Discrete FET 2N2609	2.8	0.08
Discrete bipolar 2N4045	50	0.2
Discrete MOSFET	100	0.008

from discrete components or from integrated circuits, we can model it as a noiseless source of gain together with a source of noise voltage acting in series with the input voltage, and a source of noise current flowing in parallel with the input current (Haus 1960). This is shown in Fig. 4.13.

These noise sources are white and usually assumed to be uncorrelated. They are normally specified in terms of spectral densities, $\langle e_n^2 \rangle$ in (volt)^2/Hz and $\langle i_n^2 \rangle$ in (amp)^2/Hz.

Both $\langle e_n^2 \rangle$ and $\langle i_n^2 \rangle$ are frequency dependent and manufacturers' data sheets should be consulted for full details. In general, FET input op-amps and low bias current bipolar op-amps have lower noise current while bipolar op-amps have lower noise voltages. Some typical values for the noise sources in op-amps and discrete active components are shown in Table 4.1.

The general noise model of the op-amp is shown in Fig. 4.14

The noise voltage is shown as a series voltage in the inverting terminal together with current noise generators $\langle i_{n+}^2 \rangle$ and $\langle i_{n-}^2 \rangle$ associated with the noninverting and inverting terminals respectively.

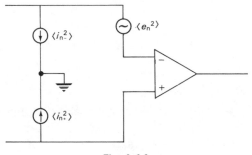

Fig. 4.14

4.5.1 Noise figure of amplifiers

It is convenient at this point to try to characterize the noise behavior of the amplifier, and a popular figure of merit used for this purpose is the noise factor F, which compares the noise performance of an amplifier with that of an ideal or noiseless amplifier.

$$F = \frac{\text{Noise power output of amplifier}}{\text{Noise power output of ideal device}}$$

The noise power output of the ideal device will be simply due to the thermal noise power of the source resistance feeding into the amplifier. Thus, we can write

$$F = \frac{\text{Noise power output of amplifier}}{\text{Power output due to source noise}}$$

Clearly, for a noiseless amplifier, $F = 1$, while, for a real amplifier, $F > 1$. In addition to the noise factor, a noise figure NF is also used. This is defined as

$$NF = 10 \log_{10} F$$

and is measured in decibels (dB).

Let us now consider several amplifiers in cascade, fed by some voltage source of resistance R_s. Each amplifier has a power gain $G_1, G_2, \ldots$, as shown in Fig. 4.15.

The input noise to the first amplifier is simply due to the source resistance R_s and is denoted by N_R. If the amplifiers were noiseless, the output noise power of the two amplifier stage would be $G_1 \times G_2 \times N_R$. However, each amplifier adds a certain noise power at each stage as shown in the figure. The overall noise factor of the two-stage amplifier is then

$$F = \frac{G_2(G_1 N_R + N_{A1}) + N_{A2}}{G_1 G_2 N_R}$$

$$= \frac{G_1 N_R + N_{A1}}{G_1 N_R} + \frac{N_{A2}}{G_1 G_2 N_R}$$

$$= \frac{G_1 N_R + N_{A1}}{G_1 N_R} + \frac{1}{G_1}\left(\frac{G_2 N_R + N_{A2}}{G_2 N_R} - 1\right)$$

$$= F_1 + \frac{F_2 - 1}{G_1}$$

where F_1 and F_2 are the noise factors of the two individual amplifiers.

We can easily extend this argument to a larger number of stages. However, the above relation for two states gives the same end result as for a larger number of stages, namely that, since G is usually very much greater than F, the major contribution to the overall noise factor of a multistage amplifier system comes from the first stage. Hence, special attention should be given to the noise in the first-stage amplifier since this will be amplified by all successive stages.

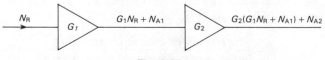

Fig. 4.15

If we know the noise figure F of the amplifier, it is possible to obtain the signal-to-noise ratio of that amplifier. Let us suppose that a voltage V_s of source resistance R_s feeds an amplifier of voltage gain A_v which has a load resistance R_L. The input noise voltage V_{ni} to the amplifier is

$$V_{ni} = \sqrt{4kTR_s\,\Delta f}$$

so that, if the total output noise voltage is V_{no}, then the noise figure of the amplifier is

$$F = \frac{V_{no}^2/R_L}{A_v^2 \cdot 4kTR_s\,\Delta f/R_L} = \frac{V_{no}^2}{4kTR_s A_v^2\,\Delta f}$$

The signal-to-noise *power* ratio at the output of the amplifier is

$$\frac{S}{N} = \frac{A_v^2 V_s^2}{V_{no}^2}$$

However,

$$V_{no}^2 = 4kTR_s A_v^2 F\,\Delta f$$

Hence

$$\frac{S}{N} = \frac{V_s^2}{4kTR_s F\,\Delta f}$$

which is the desired expression.

We can illustrate the calculation of F for a practical amplifier by considering a voltage source of resistance R_s feeding a noninverting op-amp, as shown in Fig. 4.16. Such a voltage source could be a reverse-biased photodiode or a pyroelectric detector.

The equivalent circuit for noise calculations is more complicated and is shown in Fig. 4.17.

$\langle e_n^2 \rangle$, $\langle i_{n-}^2 \rangle$, and $\langle i_{n+}^2 \rangle$ are the spectral densities of the noise associated

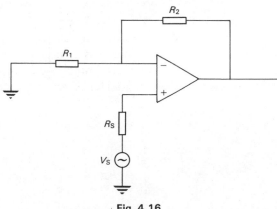

Fig. 4.16

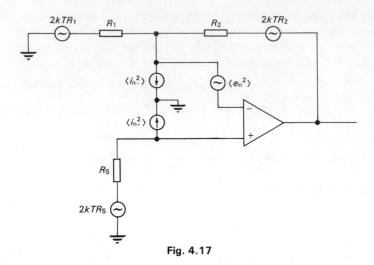

Fig. 4.17

with the amplifier. $2kTR_s$ is the spectral density of the thermal noise associated with the source resistance R_s, with similar noise voltages for R_1 and R_2. To simplify the calculations, we will assume that the infinite gain approximation applies to the op-amp.

We will calculate the mean square of the noise voltage at the output of the op-amp by assuming that the noise sources are uncorrelated and by using the superposition theorem. This theorem allows us to calculate the output noise by assuming that the total mean square output noise is the sum of the mean square output noise due to each noise source calculated independently. Practically, this means shorting out all noise voltages and open-circuiting all noise currents except the particular noise source whose effect is being calculated.

Suppose, for example, we are calculating the output noise component due to R_s. The circuit used for this calculation is shown in Fig. 4.18.

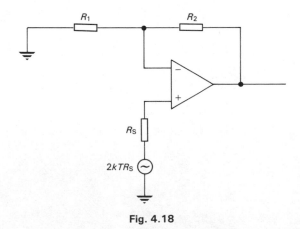

Fig. 4.18

The mean square output spectral noise density due to R_s is then

$$\langle e_{o1}^2 \rangle = \left(\frac{R_2 + R_1}{R_1}\right)^2 \cdot 2kTR_s$$

The effects of the other noise sources are easily calculated. They are presented below.

1. Due to $\langle e_n^2 \rangle$

$$\langle e_{o2}^2 \rangle = \langle e_n^2 \rangle \left(\frac{R_2 + R_1}{R_1}\right)^2$$

2. Due to $\langle i_{n+}^2 \rangle$

$$\langle e_{o3}^2 \rangle = \langle i_{n+}^2 \rangle R_s^2 \left(\frac{R_2 + R_1}{R_1}\right)^2$$

3. Due to $\langle i_{n-}^2 \rangle$

$$\langle e_{o4}^2 \rangle = \langle i_{n-}^2 \rangle (R_1 \parallel R_2)^2 \left(\frac{R_2 + R_1}{R_1}\right)^2$$

where $(R_1 \parallel R_2)$ is the parallel resistance of R_1 and R_2.

4. Due to R_1

$$\langle e_{o5}^2 \rangle = 2kT \frac{R_2^2}{R_1}$$

$$= 2kT \frac{(R_1 \parallel R_2)^2}{R_1} \left(\frac{R_2 + R_1}{R_1}\right)^2$$

5. Due to R_2

$$\langle e_{o6}^2 \rangle = 2kTR_2$$

$$= \frac{2kT}{R_2} (R_1 \parallel R_2)^2 \left(\frac{R_2 + R_1}{R_1}\right)^2$$

The total mean square spectral density at the output of the op-amp is

$$\langle e_o^2 \rangle = \left[2kTR_s + \langle e_n^2 \rangle + \langle i_{n+}^2 \rangle R_s^2 + \langle i_{n-}^2 \rangle (R_1 \parallel R_2)^2 \right.$$
$$\left. + 2kT(R_1 \parallel R_2)^2 \left(\frac{1}{R_1} + \frac{1}{R_2}\right) \right] \left(\frac{R_2 + R_1}{R_1}\right)^2$$

$$= \left[2kTR_s + \langle e_n^2 \rangle + \langle i_{n+}^2 \rangle R_s^2 + \langle i_{n-}^2 \rangle (R_1 \parallel R_2)^2 \right.$$
$$\left. + 2kT(R_1 \parallel R_2) \right] \left(\frac{R_2 + R_1}{R_1}\right)^2$$

The quantity $(R_1 + R_2)/R_1$ is the voltage gain of the amplifier. The term in the square brackets is thus the total mean square noise spectral density referred to the input of the amplifier. To obtain the total mean square noise, this expression must be integrated over the bandwidth of the amplifier

$$\langle e_{tot}^2 \rangle = \int_{-\Delta f}^{+\Delta f} \left[2kTR_s + \langle e_n^2 \rangle + \langle i_{n+}^2 \rangle R_s^2 + \langle i_{n-}^2 \rangle (R_1 \| R_2)^2 \right.$$
$$\left. + 2kT(R_1 \| R_2) \right] \left(\frac{R_2 + R_1}{R_1} \right)^2 df$$

If we ignore the frequency dependence of $\langle e_n^2 \rangle$, $\langle i_{n+}^2 \rangle$, and $\langle i_{n-}^2 \rangle$, the noise figure of the amplifier can be written

$$F = \frac{[4kTR_s + \langle e_n^2 \rangle + \langle i_{n+}^2 \rangle R_s^2 + \langle i_{n-}^2 \rangle (R_1 \| R_2)^2 + 4kT(R_1 \| R_2)] \ \Delta f}{4kTR_s \ \Delta f}$$

where we have assumed that $\langle e_n^2 \rangle$ and $\langle i_n^2 \rangle$ are defined only over positive frequencies.

We can differentiate this expression with respect to R_s to obtain a minimum value for F. This minimum occurs when

$$R_s^2 = \frac{\langle e_n^2 \rangle}{\langle i_{n+}^2 \rangle}$$

Thus, for a given source resistance, we should choose an amplifier having a given ratio of noise voltage to noise current. Integrated circuits have a ratio of noise voltage to noise current in the range 1 kΩ to 1 MΩ. Bipolar transistors range from 50 Ω to 1 MΩ, JFET discrete devices range from 1 kΩ to 1000 MΩ while discrete MOSFET devices have values between 1 MΩ and in excess of 10 000 MΩ. This latter value makes MOSFET discrete devices particularly useful when working with sensors of very high resistance.

4.6 A WORKED EXAMPLE: THE PHOTODIODE–AMPLIFIER COMBINATION (Hamstra and Wendland 1972)

In this section we shall consider, in detail, the combination of the photodiode and amplifier, and obtain expressions of the bandwidth and signal-to-noise ratio of the system. We shall consider two configurations of photodiode – photovoltaic and photoconductive – illustrated with an appropriate op-amp in Fig. 4.19. In the photovoltaic system, the photocurrent is converted to a voltage by a transimpedance (current to voltage) amplifier, while in the photoconductive case the photocurrent is used to develop a voltage across a load resistor, which is then amplified by a voltage amplifier.

In Fig. 4.19(a), C_f represents the inevitable stray capacitance associated with the feedback resistor R_f. In (b), R_B is a bias resistor and R_L the load resistor across which the photoinduced voltage is developed. (As we shall see later, R_B adds noise to the system, and in most practical systems is set equal to zero.) In the following calculations, the amplifier will be considered

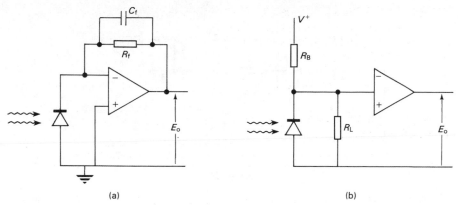

(a)   (b)

Fig. 4.19

to have an infinite input resistance, an input capacitance C_1, and a frequency-dependent open-loop gain

$$A(\omega) = \frac{A(0)}{1 + j \dfrac{\omega}{\omega_1} A(0)}, \qquad A(0) \gg 1$$

where ω_1 represents the unity gain bandwidth.

From Chapter 2, we should recall that the photodiode can be modeled as a current generator across which there is a shunt resistance r_d and capacitance C_d associated with the junction, together with a small series resistance R_s. This is shown in Fig. 4.20 for reference.

In the photovoltaic mode, r_d is smaller than in the photoconductive mode, while C_d is smaller in the photoconductive mode. Implicit in the following calculations is that the frequency response of the photodiode is determined by the RC time constant effect of the photodiode.

Let us now consider the photovoltaic system in detail. The equivalent circuit for bandwidth calculations is presented in Fig. 4.21. $\langle I_p \rangle$ is the photocurrent generated by the illuminated diode.

Assuming that the input resistance of the amplifier is infinite, then equating currents at the summing junction,

$$i_1 + i_f = 0$$

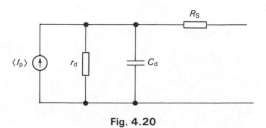

Fig. 4.20

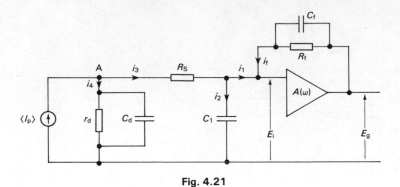

Fig. 4.21

Now

$$i_3 = i_1 + i_2$$

so that

$$\langle I_p \rangle = i_3 + i_4$$

If E_i is the voltage into the amplifier

$$i_2 = j\omega C_1 E_i$$

and therefore

$$i_3 = +j\omega C_1 E_i - i_f$$

Further,

$$V_A - E_i = i_3 R_s$$

or

$$i_4 \left[\frac{r_d}{1 + j\omega C_d r_d} \right] = E_i + i_3 R_s$$

From these expressions, it follows that

$$\langle I_p \rangle = \frac{(1 + j\omega C_d r_d)}{r_d} \, [E_i(1 - j\omega C_1 R_s) - i_f R_s] + j\omega C_1 E_i - i_f$$

where

$$i_f = (E_o - E_1) \frac{(1 + j\omega C_f R_f)}{R_f}$$

R_s is usually very small and it can be ignored. Then

$$\langle I_p \rangle \approx \frac{(1 + j\omega C_d r_d)}{r_d} \, E_i + j\omega C_1 E_i - i_f$$

$$= E_i \left(\frac{1}{r_d} + \frac{1}{R_f} + j\omega [C_1 + C_d + C_f] \right) - E_o \left(\frac{1}{R_f} + j\omega C_f \right)$$

We can substitute for E_i since

$$\frac{E_o}{E_i} = \frac{-A(0)}{1 + j\dfrac{\omega}{\omega_1} A(0)}$$

and it follows that

$$\langle I_p \rangle = -E_o\left[\frac{1}{A(0)r_d} + \frac{1}{R_f}\left(\frac{1}{A(0)} + 1\right) + j\omega C_f\left(\frac{1}{A(0)} + 1\right)\right.$$

$$\left. + \frac{j\omega}{A(0)}(C_1 + C_d) + j\frac{\omega}{\omega_1}\left(\frac{1}{r_d} + \frac{1}{R_f}\right) - \frac{\omega^2}{\omega_1}(C_1 + C_d + C_f)\right]$$

This expression can be simplified considerably by noting that $A(0) \gg 1$. Then

$$\langle I_p \rangle \approx -E_o\left[\frac{1}{R_f} + j\omega C_f + \frac{j\omega}{A(0)}(C_1 + C_d)\right.$$

$$\left. + \frac{j\omega}{\omega_1}\left(\frac{1}{r_d} + \frac{1}{R_f}\right) - \frac{\omega^2}{\omega_1}(C_1 + C_d + C_f)\right]$$

This expression for the output voltage E_o illustrates the phenomenon of gain peaking. This is due to the factor

$$\frac{1}{R_f} - \frac{\omega^2}{\omega_1}(C_1 + C_d + C_f)$$

If this goes to zero before the complex term causes E_o to fall away, then the output will peak in the manner illustrated in Fig. 4.22. Such an effect causes instability, the major problem in this configuration.

Gain peaking will occur if

$$\sqrt{\frac{\omega_1}{R_f(C_1 + C_d + C_f)}} < \frac{1}{R_f C_f + \dfrac{R_f}{A(0)}(C_1 + C_d) + \dfrac{1}{\omega_1}\left(\dfrac{R_f}{r_d} + 1\right)}$$

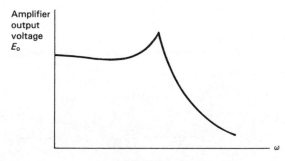

Fig. 4.22

Let us take as typical values $R_f = 10 \text{ M}\Omega$, $A(0) = 100$, $r_d = 1 \text{ G}\Omega$, $\omega_1 = 10^4 \text{ Hz}$, $C_1 = 2 \text{ pF}$, $C_d = 2 \text{ pF}$, $C_f = 5 \text{ pF}$. Then

$$\sqrt{\frac{\omega_1}{R_f(C_1 + C_d + C_f)}} = 10^4$$

and

$$\frac{1}{R_f C_f + \dfrac{R_f}{A(0)}(C_1 + C_d) + \dfrac{1}{\omega_1}\left(\dfrac{R_c}{r_d} + 1\right)} = 2 \times 10^4$$

so that gain peaking will occur. By adding an external capacitor across R_f, say 100 pF, gain peaking can be avoided for this particular amplifier.

In practice, ω_1 and $A(0)$ are much larger than the values assumed here. For example, for the 741 op-amp, ω_1 has a typical value of 1.5 MHz and so we can ignore terms in $1/\omega_1$. Then

$$E_o = \frac{-\langle I_p \rangle R_f}{1 + \dfrac{j\omega}{A(0)}(C_1 + C_d + A(0)C_f)R_f}$$

The bandwidth is thus

$$\Delta f = \frac{A(0)}{2\pi R_f(C_1 + C_d + A(0)C_f)}$$

If signals of frequency 0 to f_{max} are encountered, the output of the circuit is independent of frequency if

$$\frac{2\pi}{A(0)} f_{max} R_f(C_1 + C_d + A(0)C_f) \ll 1$$

or

$$f_{max} \ll \frac{A(0)}{2\pi R_f(C_1 + C_2 + A(0)C_f)}$$

This simplified expression shows that the capacitance C_f will limit the bandwidth of the system. Its effect can be canceled by means of a compensation network, as shown in Fig. 4.23.

If R_1 is kept very much smaller than R_f, then it will contribute negligible extra noise to the system.

In a typical case, $R_f = 10 \text{ M}\Omega$, $C_f = 5 \text{ pF}$, $C_1 = 2 \text{ pF}$, $C_2 = 2 \text{ pF}$, $A(0) = 100$. Without compensation,

$$\Delta f = 3 \text{ kHz}$$

With compensation,

$$\Delta f = 200 \text{ kHz}$$

which is a significant improvement in bandwidth.

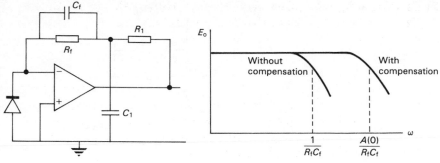

Fig. 4.23

The noise model of the system is more complex (Fig. 4.24).

$\sqrt{\langle e_a^2 \rangle}$, $\sqrt{\langle i_n^2 \rangle}$ represent the noise voltage and noise current of the amplifier respectively. The term eI_B represents the spectral density of the shot noise due to background photons and $2kT/r_d$ is the spectral density of the thermal noise associated with the slope resistance of the diode. These latter noise sources can be lumped together with the noise current of the amplifier to give a single noise current source $\sqrt{\langle i_{tot}^2 \rangle}$. We again assume that the input resistance of the amplifier is infinite and the equivalent noise model of the photodiode–op-amp configuration is now as shown in Fig. 4.25.

The spectral density of the noise at the output due to the lumped current source $\sqrt{\langle i_{tot}^2 \rangle}$ is

$$\langle e_{n1}^2 \rangle = \frac{-R_f}{1 + \dfrac{j\omega}{A(0)}(C_1 + C_d + A(0)C_f)}^{2} \langle i_{tot}^2 \rangle$$

The noise current due to the feedback resistor produces a noise voltage

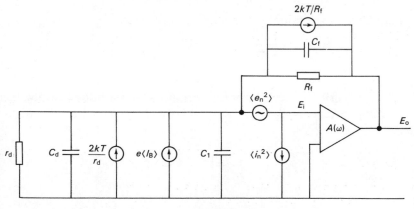

Fig. 4.24

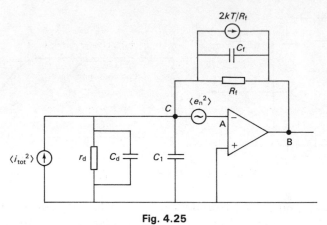

Fig. 4.25

across the feedback resistor R_f of spectral density

$$\langle e_{R_f}^2 \rangle = \left(\frac{2kT}{R_f}\right) \frac{1}{\left(\dfrac{1}{R_f} + j\omega C_f\right)^2}$$

The noise voltage due to this resistor, at the output of the amplifier ($\langle e_{n2}^2 \rangle$) is thus

$$\sqrt{\langle e_{n2}^2 \rangle} - \sqrt{\langle e_A^2 \rangle} = \left(\frac{2kT}{R_f}\right)^{1/2} \frac{1}{\left(\dfrac{1}{R_f} + j\omega C_f\right)}$$

where $\langle e_A^2 \rangle$ is the noise voltage at the input A to the amplifier due to R_f. Writing

$$\sqrt{\langle e_{n2}^2 \rangle} = -A\sqrt{\langle e_A^2 \rangle}$$

then

$$\langle e_{n2}^2 \rangle = \left(\frac{2kT}{R_f}\right) \frac{1}{\left(\dfrac{1}{R_f} + j\omega C_f\right)^2} \cdot \frac{1}{\left(1 + \dfrac{1}{A}\right)^2}$$

The spectral density of the noise at the output $\langle e_{n3}^2 \rangle$ due to $\langle e_n^2 \rangle$ is obtained by equating currents at the summing junction C to the amplifier. In this case,

$$\left(\sqrt{\langle e_{n3}^2 \rangle} - \sqrt{\langle e_A^2 \rangle} - \sqrt{\langle e_n^2 \rangle}\right)\left(\frac{1}{R_f} + j\omega C_f\right) = \left(\sqrt{\langle e_n^2 \rangle} + \sqrt{\langle e_A^2 \rangle}\right)j\omega C_1$$

$$+ \left(\sqrt{\langle e_A^2 \rangle} + \sqrt{\langle e_n^2 \rangle}\right)\left(\frac{1}{r_d} + j\omega C_d\right)$$

so that

$$\langle e_{n3}^2 \rangle = \langle e_n^2 \rangle \left\{ \frac{\frac{1}{r_d} + \frac{1}{R_f} + j\omega(C_d + C_f + C_1)}{\frac{1}{R_f} + \frac{1}{A}\left(\frac{1}{r_d} + \frac{1}{R_f}\right) + j\omega C_f + \frac{j\omega}{A}(C_1 + C_d A C_f)} \right\}^2$$

The total noise power at the output of the amplifier is obtained by adding the square of the spectral densities of the three output noise expressions and integrating over the bandwidth of interest.

$$\langle V_{no}^2 \rangle = \int_{-\Delta f}^{+\Delta f} [\langle e_{n1}^2 \rangle + \langle e_{n2}^2 \rangle + \langle e_{n3}^2 \rangle] \, df$$

A simple (and more tractable) result can be obtained by assuming that $A \gg 1$ and that $C_f = 0$ (i.e. compensation has been achieved). Then

$$\langle v_{no}^2 \rangle = \int_{-\Delta f}^{+\Delta f} \left[R_f^2 \langle i_{tot}^2 \rangle + 2kTR_f + \left\{ \left(1 + \frac{R_f}{r_d}\right)^2 + \omega^2 R_f^2 (C_1 + C_d)^2 \right\} \langle e_n^2 \rangle \right] \, df$$

$$= \left[2R_f^2 \langle i_{tot}^2 \rangle + 4kTR_f + 2 \left\{ \left(1 + \frac{R_f}{r_d}\right)^2 + \frac{4\pi^2}{3}(C_1 + C_d)^2 R_f^2 (\Delta f)^2 \right\} \langle e_n^2 \rangle \right] \Delta f$$

This expression illustrates the need to use low noise amplifiers with low values of input capacitance. In particular, it implies the use of amplifiers having FET input stages. Indeed, hybrid integration of p-i-n photodiodes and FET amplifiers are now available (the so-called PINFET). Both silicon and gallium arsenide diodes have been integrated in this way. The FETs used are gallium arsenide MESFETs which have extremely low input capacitance and very low gate leakage current. These devices find their main application in fiber optic communication systems where they have been challenging the sensitivity of avalanche photodiodes (Siegel and Channin 1984).

For the photoconductive system, the calculations are less complicated. The equivalent circuit for bandwidth calculations is shown in Fig. 4.26.

For simplicity, we shall ignore R_s. The voltage generated at the input of the voltage amplifier is

$$E_i = i_d Z$$

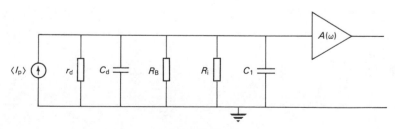

Fig. 4.26

where Z is the parallel combination of the resistances and capacitances

$$\frac{1}{Z} = \frac{1}{r_d} + \frac{1}{R_B} + \frac{1}{R_L} + j\omega(C_1 + C_d) = \frac{1}{R_{||}} + j\omega C_t$$

Thus

$$E_i = \frac{i_d R_{||}}{1 + j\omega C_t R_{||}}$$

so that

$$E_o = i_d \left(\frac{R_{||}}{1 + j\omega C_t R_{||}}\right)\left(\frac{A(0)}{1 + j\,\dfrac{\omega}{\omega_1}\,A(0)}\right)$$

For the photodiode in the photoconductive mode, $r_d \gg R_B, R_L$ and, with values of $A(0)$ and ω_1 obtainable with op-amps, the bandwidth of the system is approximately

$$\Delta f = \frac{1}{2\pi(R_B + R_L)C_t}$$

With $R_B = 0$, it can be seen that the bandwidth in this case is inversely proportional to R_L.

The model for the signal-to-noise calculations is presented in Fig. 4.27. In the back-biased photoconductive mode, we saw in Chapter 2 that the expression for the shot noise is different from that in the zero bias case. In the figure, the spectral density of the shot noise due to the signal, the background, and the dark current is lumped together as $\langle i_{sh}^2 \rangle$ where

$$\langle i_{sh}^2 \rangle = e(\langle I_p \rangle + \langle I_B \rangle + \langle I_d \rangle)$$

The spectral density of the total noise current entering the amplifier is

$$\langle i_{tot}^2 \rangle = 2kT\left(\frac{1}{R_B} + \frac{1}{R_L}\right) + \langle i_{sh}^2 \rangle + \langle i_n^2 \rangle + \frac{\langle e_n^2 \rangle}{Z_A^2}$$

where

$$\frac{1}{Z_A} = \frac{1}{R_B} + \frac{1}{R_L} + j\omega C_t$$

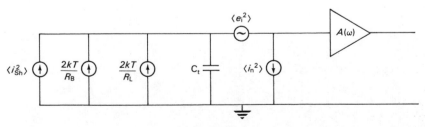

Fig. 4.27

The signal-to-noise ratio can now be written down, but let us assume, for the sake of simplicity, that $R_B = 0$ and that we are working at low frequencies where

$$\omega \ll \frac{1}{R_{\|}C_t}$$

$$\frac{S}{N} = \frac{I_p R_L}{[(\langle i_{sh}^2 \rangle + \langle i_n^2 \rangle)R_L^2 + \langle e_n^2 \rangle + 4kTR_L]^{\frac{1}{2}} \Delta f^{\frac{1}{2}}}$$

4.7 FILTERS

The signal-to-noise ratio at the output of the combined transducer–amplifier system depends on the bandwidth of the measurement system. If this bandwidth could be limited to allow through only the frequencies of interest, then an improvement in the signal-to-noise ratio will occur.

The circuit which accomplishes this function is known as a filter. A filter will attenuate some frequencies more than others. Four types of filter in general use as shown in Fig. 4.28.

Probably the best known is the low pass RC filter, illustrated in Fig. 4.29. The output voltage of this filter is

$$V_{out} = \frac{1}{\sqrt{1 + \omega^2 C^2 R^2}} V_{in}$$

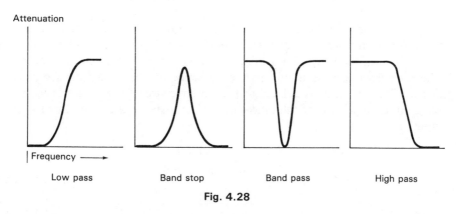

Attenuation

| Frequency ⟶

Low pass   Band stop   Band pass   High pass

Fig. 4.28

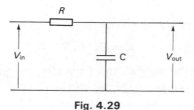

Fig. 4.29

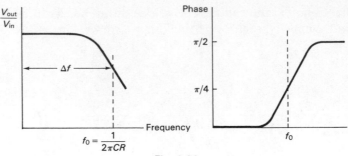

Fig. 4.30

When $\omega CR \ll 1$, the output voltage is not attenuated, but when $\omega CR \gg 1$, the output voltage is attenuated more and more as the frequency ω increases. The output voltage has fallen by a factor of $1/\sqrt{2}$ at a frequency

$$\omega_0 = \frac{1}{CR}$$

The response of this filter is shown in Fig. 4.30.

The frequency f_0 is known as the 3 db point, and we can define the 3 db bandwidth of this filter as

$$\Delta f = f_0 - 0 = \frac{1}{2\pi CR}$$

Above f_0, the response of the filter falls off quite slowly at 6 db per octave.

The bandwidth of the filter is not the only important parameter. The time taken for the output to reach a specified fraction of its final value when the input changes suddenly (the rise time of the filter) is also a relevant factor. The nonzero rise time is due to the presence of the capacitor. Conventionally, this rise time is defined as the time required for the output to rise from 10 to 90 percent of the final value when a step pulse is applied to the input. If

$$V_{in} = 0 \qquad \text{for } t < 0$$
$$= A \qquad \text{for } t \geqslant 0$$

then

$$v_{out} = A(1 - e^{-t/RC})$$

and the risetime can easily be shown to be

$$\tau_r = RC \ln 9$$

Thus, as we reduce the bandwidth of the filter (by increasing the product RC), we also slow down the response of the filter to rapid changes in the input signal.

Let us see the effect of this *RC* filter on white noise. This white noise has spectral density

$$G(f) = \frac{n_o}{2}$$

If this is applied to the input of the filter, the output noise spectral density is (remember spectral density is a measure of noise *power*):

$$G_o(f) = \frac{n_0/2}{1 + (\omega CR)^2} = \frac{n_0/2}{1 + (f/f_o)^2}$$

and so the average noise power at the output is

$$N_o = \int_{-\infty}^{+\infty} G(f) \, df = \int_{-\infty}^{+\infty} \frac{n_0/2 \, df}{1 + (f/f_o)^2} = \frac{n_o \pi}{2} f_0$$

$$= \frac{n_o \pi}{2} \Delta f$$

where Δf is the 3 dB bandwidth of the filter. Thus, for white noise at least, we can reduce the amount of noise at the output by decreasing the bandwidth.

The small signal closed-loop (3 dB) bandwidth is not the only definition of bandwidth and, when dealing with the noise-rejecting properties of filters, it is usual to employ another definition, the *noise equivalent bandwidth B*. To arrive at a mathematical expression for *B*, the low pass filter characteristic is replaced with an ideal rectangular filter function, as shown in Fig. 4.31.

If the filter has a frequency response $A(\omega)$, then

$$v_{out} = A(\omega)v_{in}$$

while the ideal filter has a frequency response

$$A(\omega) = A(0) \qquad -B \leqslant \omega \leqslant +B$$
$$= \quad 0 \qquad \text{otherwise}$$

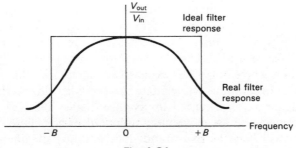

Fig. 4.31

Now consider white noise applied to the input of these filters. Then

$$\text{Noise } \textit{power} \text{ output (ideal filter)} = \frac{n_0}{2} |A(0)|^2 \, 2B$$

$$= n_o B |A(0)|^2$$

$$\text{Noise } \textit{power} \text{ output (real filter)} = \frac{n_0}{2} \int_{-\infty}^{+\infty} |A(\omega)|^2 \, d\omega$$

$$= n_o \int_0^\infty |A(\omega)|^2 \, d\omega$$

The noise equivalent bandwidth B is obtained by equating these two expressions

$$B = \frac{1}{|A(0)|^2} \int_0^\infty |A(\omega)| \, d\omega$$

There will, of course, be a relationship between B and 3 dB bandwidth. For the simple low pass RC filter,

$$B = \frac{1}{4RC}$$

Thus, for any filter,

$$\text{Noise power output} = n_o B |A(0)|^2$$

which is proportional to the bandwidth B of the filter. To decrease this noise output, the bandwidth of the filter should be made as narrow as possible to attenuate the noise, assuming that the bandwidth of the noise is very much greater than the filter bandwidth. This cannot go on without limit since the response time of the system increases, so that the limitation on the bandwidth is set by the rate of variation of the signal itself.

4.7.1 Active filters

The previous discussion of filters has assumed that they are constructed entirely out of passive components. However, active filters which use operational amplifiers in conjunction with resistor–capacitor networks promise better performance at the relatively low frequencies used in signal processing applications where bulky and expensive inductors would be required in a passive filter. In addition, active filters can provide gain together with high input impedance and low output impedance (Sallen and Key 1955).

The performance of an active filter is limited principally by its op-amp. The op-amp will have its own bandwidth and so will limit the highest frequency at which the active filter will operate. A high gain–bandwidth

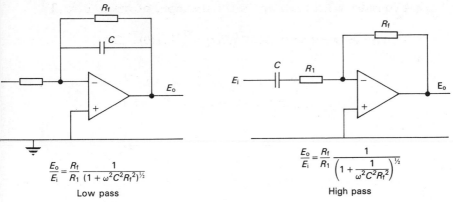

$$\frac{E_o}{E_i} = \frac{R_f}{R_1} \frac{1}{(1 + \omega^2 C^2 R_f^2)^{\frac{1}{2}}}$$

Low pass

$$\frac{E_o}{E_i} = \frac{R_f}{R_1} \frac{1}{\left(1 + \dfrac{1}{\omega^2 C^2 R_f^2}\right)^{\frac{1}{2}}}$$

High pass

Fig. 4.32

product op-amp such as a bi-FET op-amp (mixed bipolar and FET IC) is essential. In addition, the op-amp should have a high slew rate since, if it becomes slew rate limited, the op-amp will generate harmonics at its output. Even at signal processing frequencies, passive filters cannot be ruled out for certain critical applications where the theoretical limits of the performance of a system are determined by the filter. For example, if noise must be kept to an absolute minimum, or offset is undesirable, a passive filter would be unavoidable.

Simple low pass and high pass filters are shown in Fig. 4.32 The roll-off of these filters is 6 db per octave, which is generally not fast enough. Filters are characterized by their roll-off, and an nth-order filter has a roll-off of $6n$ db per octave, making the above filters first order. There are several commonly used higher order active filters. The most common is the unity gain filter, of which a second-order low pass version is shown in Fig. 4.33.

The analysis of this circuit is quite straightforward using the rules for the infinite gain amplifier, viz. the $+$ and $-$ terminals are at the same potential and no current flows into the terminals. Then

$$E_B = E_o$$

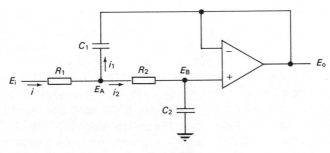

Fig. 4.33

and

$$i = i_1 + i_2$$

Thus

$$\frac{E_i - E_A}{R_1} = \frac{E_A - E_o}{R_2} + \frac{E_A - E_o}{1/j\omega C_1}$$

and

$$\frac{E_A - E_o}{R_2} = \frac{E_o}{1/j\omega C_2}$$

from which equations

$$\frac{E_o}{E_i} = \frac{1}{[(1 - \omega^2 C_1 C_2 R_1 R_2 + j\omega C_2(R_1 + R_2))]}$$

The filter response function has now been obtained and the detailed response of the system depends on the values of the components. It is advantageous to tailor the filter response function to correspond to certain mathematical functions. Since instrumentation systems are concerned with measurements of amplitudes, the configuration usually employed is the Butterworth filter, which has a response

$$|A(\omega)|^2 = \frac{1}{1 + (\omega/\omega_c)^{2n}}$$

where ω_c is the 3 dB cut-off frequency. This filter has a very flat amplitude response in the pass band but has a nonlinear phase response. For the above filter, a second-order Butterworth filter is obtained if

$$R_1 = R_2, \; C_1 = 2C_2$$

and the filter has a 3 db cut-off of

$$f_c = \frac{1}{2\pi R_1 \sqrt{C_1 C_2}}$$

For $R_1 = 10 \, \text{k}\Omega$ and $C_1 = 220 \, \text{pF}$, a cut-off frequency of 100 kHz is obtained.

There are several advantages of these unity gain filters. The low pass filter can be converted into a high pass filter by interchanging the resistors and capacitors. Higher order filters can be realized simply by cascading several such filters. They involve a small number of components and so are relatively low cost. They are, however, relatively sensitive to small changes in component values.

Another common family of active filter is the multiple feedback filter. They are not as easily tuned as the unity gain variety and the gain depends on component ratios. Such filters are useful for fixed parameter (e.g. constant frequency) filters. An example of a second-order Butterworth

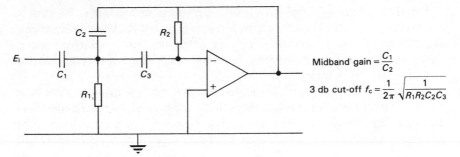

Midband gain $= \dfrac{C_1}{C_2}$

3 db cut-off $f_c = \dfrac{1}{2\pi} \sqrt{\dfrac{1}{R_1 R_2 C_2 C_3}}$

Fig. 4.34

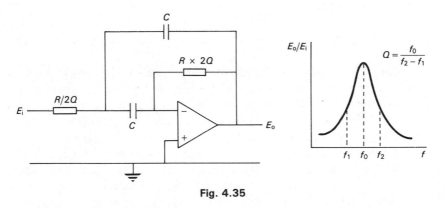

$Q = \dfrac{f_0}{f_2 - f_1}$

Fig. 4.35

multiple feedback filter is shown in Fig. 4.34, and a band pass filter in Fig. 4.35.

The maximum Q value for the circuit in Fig. 4.35 is low (e.g. 20 at 1 kHz) since a high Q circuit becomes unstable and tends to oscillate.

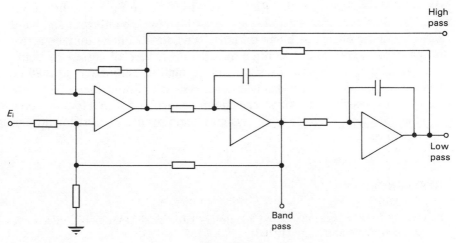

Fig. 4.36

The three types of filter (low pass, high pass and band pass) can be achieved by the state variable active filter (Fig. 4.36). It allows independent adjustment of centre frequency and the Q-factor, and is relatively insensitive to variations in component values so that Q values of several hundred at 1 kHz can be obtained.

The major application of op-amps to signal processing have now been covered. Other signal processing tasks that have previously been undertaken using analog circuitry, such as linearizing the output of a transducer, have not been covered here since these tasks are now being performed more and more with microprocessors. These devices are having a profound effect on signal processing and will be covered in a later chapter.

SUMMARY

In this chapter we have considered some aspects of signal conditioning in which the experimenter is concerned with the quality of the signal presented by the transducer. This was considered as distinct from the concept of signal processing, which is discussed in the next chapter. Two aspects of signal conditioning have been considered – amplification and filtering. After a discussion of the ideal op-amp and some of the more important configurations, the limitations of amplifiers, such as bandwidth, slew rate, offset voltages and currents, and drift characteristics, were introduced, the latter factors being the major selection factors in the majority of amplifier applications. For very low level signals, the limit on the system resolution is determined by the noise input to the amplifier and the noise generated in the amplifier. A noise model of the amplifier, consisting of a noise voltage and noise current together with a noiseless amplifier, was developed, and the concept of the noise factor of an amplifier introduced. The final sections of the chapter dealt with the use of filters to improve signal quality. Four types of filter were encountered – low pass, high pass, band stop, and band pass – and the effects of filters on noise considered. Filters may be active or passive. Active filters, containing op-amps, were discussed in detail because of their ability to provide low cost, high stability filters at the low frequencies employed in instrumentation systems without the need to use bulky and expensive inductors. The chapter concluded with a look at some of the practical active filter systems employed in instrumentation applications.

REFERENCES

Graeme, J. 1980. Instrumentation amplifiers sift signal from noise. *Electronic Design*, September, pp. 119–23.

Hamstra, R.H., and Wendland, P. 1972. Noise and frequency response of silicon photodiode operational amplifier combination. *Applied Optics*, Vol. 11, pp. 1539–47.

Haus, H.A. 1960. Representation of noise in linear twoports. *Proceedings of the IRE*, Vol. 48, pp. 69–74.

Jaquay, J.W. 1973. Designers' guide to instrumentation amplifiers. *Electronic Design News*, May.

Letzer, S., and Webster, N. 1970. Noise in amplifiers. *IEEE Spectrum*, August, pp. 69–75.

Sallen, R.P., and Key, E.L. 1955. A practical method of designing RC active filters. *IRE Transactions on Circuit Theory, CT-2*, March.

Schlick, L.L. 1971. Linear circuit applications of operational amplifiers. *IEEE Spectrum*, April, pp. 36–50.

PROBLEMS

4.1 Calculate the output voltage error in the following circuit for:
(a) the 741 (bipolar) op-amp,
(b) the CA3140 (FET) op-amp.

	741	3140
Input offset voltage	1 mV	5 mV
Input offset current	20 nA	0.5 pA
Input bias current	80 nA	10 pA
Input offset voltage temperature coefficient	5 μV/$^\circ$C	8 μV/$^\circ$C
Input offset current temperature coefficient	0.5 nA/$^\circ$C	0.025 pA/$^\circ$C

If the temperature changes by 10°, calculate the new output voltage error.

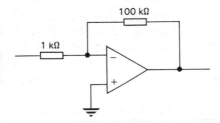

4.2 A 741 op-amp has a slew rate of 0.5 V/μs. Calculate the maximum frequency of the sine wave input to the 741 which will produce an undistorted output, if the input has a peak amplitude of (a) 1 mV, (b) 10 V.

4.3 A differential signal of magnitude 1 mV is applied to the following differential amplifier, in the presence of 1 V common mode signal. If the common mode rejection ratio of the op-amp is 90 dB, calculate the output voltage signal-to-noise ratio.

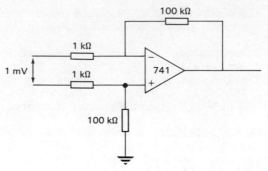

4.4 A voltage source of 1 mV amplitude and resistance 1 kΩ feeds a voltage amplifier of gain 100 with noise voltage and current:

$$\langle v^2 \rangle = 100 \ (\text{nV})^2/\text{Hz}$$
$$\langle i^2 \rangle = 0.36 \ (\text{pA})^2/\text{Hz}$$

Calculate the signal-to-noise ratio at the output of the amplifier in a bandwidth of 10 kHz.

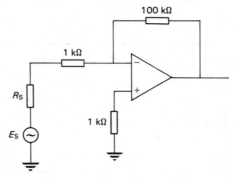

4.5 Calculate (a) the cut-off frequency, and (b) the roll-off, for the low pass Butterworth filter shown in the diagram.

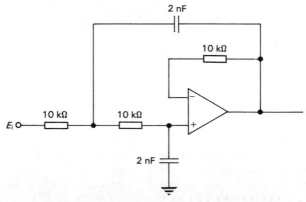

4.6 Show that the noise equivalent bandwidth B of a simple RC low pass filter is

$$B = 1/4RC$$

5

Signal Processing

Previously we have dealt with the transduction of optical signals together with some signal conditioning, which aims at improving the signal quality. We are now going to look at the task of signal processing, which involves manipulation of that signal in some way. This chapter will deal with some of the more conventional tools used in this area, such as the phase sensitive detector and boxcar integrator, while leaving the newer techniques based on microprocessors to Chapter 6. All the techniques discussed here have in fact been implemented digitally in some form or other, and those implementations will be discussed in the relevant chapter. However, the analog variants of these instruments still abound in laboratories and this justifies a discussion of their operation.

The particular signal processing technique we use will depend very much on the transducer we are employing, and the form and magnitude of the photon flux. Remembering that the light incident on the transducer is in the form of individual photon events, then, with a (hypothetically) infinitely fast detector and electronics, we should observe a series of pulses as shown in Fig. 5.1.

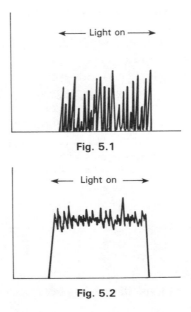

Fig. 5.1

Fig. 5.2

However, for many detectors and for a rate of photon arrival much faster than the resolution time of the subsequent electronics, what we will detect is a d.c. signal with noise superimposed on it, as shown in Fig. 5.2. Furthermore, the light may be modulated in some way; for example, chopped or pulsed light may be incident on the detector. All these considerations will determine to some extent the type of signal processing to be employed.

5.1 PHOTON COUNTING

Photon counting is a technique that has been employed for some years with photomultipliers when the light intensity is low. The best electronic amplifiers can resolve pulses arriving at a rate of about 10^9 per second. If these pulses are produced by a photomultiplier which has light of wavelength 550 nm incident on it, then the maximum light intensity at which we can resolve the individual pulses is approximately 0.3×10^{-9} W. At higher intensitities, photons will be arriving during the 'dead' time of the electronics and will not be registered. At lower intensities, we are now able to resolve the individual pulses due to the arrival of photons, and the greatest amount of information about the light is obtained by counting the photoelectron pulses. The inherent advantages of photon counting over rival techniques are due to the counting process itself. It is a digital process and can be used directly with digital microprocessor systems. Since it is digital, it is less susceptible to long term drift and $1/f$ noise associated with analog systems. Also, it is possible to discriminate against pulses which have a low probability of coming from the incident photon flux. This makes photon counting less sensitive to sources of noise other than thermionic emission (Amoss and Davidson 1972, Foord et al. 1969, James 1967, Jones et al. 1971, Oliver and Pike 1968).

The basic photon counting system is shown in Fig. 5.3. Photons arrive at the photomultiplier cathode and give rise to a burst of photoelectrons at the anode of the photomultiplier as well as electrons arising from various processes within the photomultiplier (dark current). These charge pulses are amplified and passed to the discriminator, which will reject all pulses below a certain height and accept the rest, whose rate can be counted by the digital counter. Let us look at these components to see how they affect the overall design of the photon counting system.

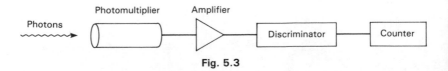

Fig. 5.3

(a) Photomultiplier
The operation of the photomultiplier has been discussed in Chapter 2 on

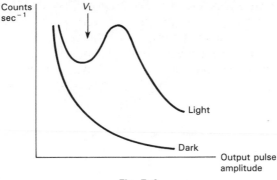

Fig. 5.4

optical transducers. We will discuss here only the factors which contribute to the dark noise of the photomultiplier.

A number of physical processes are responsible for the dark current in photomultipliers. The most fundamental process is thermionic emission from the photocathode, which is temperature dependent. A second process arises from ions from adsorbed gases on the photomultiplier envelope and on metal electrodes. Large noise pulses may also be generated by cosmic rays and background radioactivity. Finally, low level noise is generated by the statistical nature of the secondary emission in the multiplication process.

The distribution of voltage pulse heights proportional to the charge in each pulse of a typical photomultiplier is shown in Fig. 5.4.

When there is no light incident on the photomultiplier, most pulses are relatively low since most thermionic emission originates from the dynodes. The large pulses are due to cosmic rays and residual radioactivity in the glass envelope of the photomultiplier. In the presence of light, a pronounced increase in higher pulses occurs. The discriminator allows only voltage pulses above a certain level V_L to be accepted, thus rejecting a large proportion of the dark count; the level V_L is set to maximize the light-count to dark-count ratio. Careful design of modern photomultipliers has led to a very small count due to residual activity, etc. so that an upper discriminator level to reject these pulses is not necessary.

In photomultipliers that incorporate S20 and NEA cathodes, temperature-dependent thermionic emission from the photocathode makes the largest contribution to the dark current at room temperatures. This thermionic current will fall as the temperature is lowered below room temperature in line with Richardson's law, but the dark current asymptotically approaches a minimum at $\sim -30°C$ as shown in Fig. 5.5 (Morton 1968).

Nothing is gained by cooling below this sort of temperature and it is now common practice to operate red-sensitive photomultipliers with a greatly reduced photocathode area and a correspondingly low (< 100 counts s^{-1})

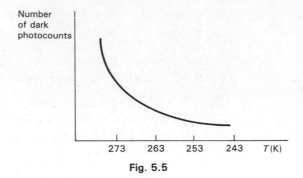

Fig. 5.5

dark-count rate at room temperature. The residual dark current below $T = -30°C$ is thought to be due to nonthermal radioactivity in the photocathode or envelope material and to cosmic radiation.

(b) Amplifier

The amplifier is used to amplify the pulses at the output of the photomultiplier tube to a level compatible with the operating range of the discriminator (50 mV to 1 V). The amplifier should have a fast rise time with low input noise and good linearity to preserve the pulse-height characteristics of the photomultiplier. Two amplifier types can be used (Fig. 5.6). The first is a voltage sensitive amplifier, where the current pulse from the photomultiplier is converted into a voltage pulse and then amplified. The second is a charge sensitive type which uses a capacitor to integrate the charge pulse; the output voltage of the amplifier is then proportional to the charge in the pulse. The charge sensitive amplifier is less sensitive to capacitively coupled noise and strongly rejects background such as r.f. noise.

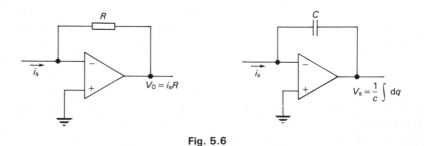

Fig. 5.6

(c) Discriminator

Modern discriminator systems use fast sense amplifiers which have a pulse pair resolution time in the nanosecond region. These are free of regenerative effects which result in multiple pulse responses to a single input event.

The effect of the pulse pair resolution time of the electronics can be seen by a simple calculation. Suppose there are N_0 pulses per second arriving at the electronics. If we now count these pulses over a time T, we would expect N_0T pulses at the output of the discriminator. However, every time a pulse is detected by the electronics there will be 'dead' time t_d during which the system will not accept pulses. During the measurement time T, the total dead time will be $NT \times t_d$, so that the actual time the system can accept pulses during the measurement time T is

$$T_a = T(1 - N_0 t_d)$$

The ratio T_a/T is obviously the fraction of the pulses at the input to the electronics appearing at the output of the discriminator and so will reflect the counting accuracy of the system.

Suppose we have a count rate of 10^6 count s^{-1} and a dead time of 10 ns, then

$$\frac{T_a}{T} = 1 - 10^6 \cdot 10^{-8} = 0.99$$

which is a 99 percent counting accuracy.

If N_0 rises, say to 10^7 counts per second, this accuracy falls to 90 percent. Similarly, the effects of lower pulse pair resolution time t_d can be seen. For $N_0 = 10^6$ counts per second and $t_d = 100$ ns, we are down to 90 percent counting accuracy.

5.1.1. Modes of operation

Direct counting

In the simplest method of photon counting, the number of photomultiplier pulses with pulse heights between the two discriminator levels are counted during a predetermined interval. This number will contain a component due to the signal photons and another due to the dark current. If there are N_s pulses per second due to signal photons and N_d pulses per second due to dark current, then, in an interval T, there will be an estimated signal count of

$$[(N_s + N_d) - N_d]T$$

and, assuming Poisson statistics, the r.m.s. error in the number counted by the system is:

$$\sqrt{(\Delta N)^2} = (N_s + N_d)^{1/2} T^{1/2}$$

The signal-to-noise ratio is

$$\frac{S}{N} = \frac{N_s T}{(N_s + N_d)^{1/2} T^{1/2}} = \frac{N_s}{(N_s + N_d)^{1/2}} \cdot T^{1/2}$$

Thus, by counting for a longer period, an improvement in the signal-to-noise ratio will occur.

Several experimental precautions should be taken in low level counting systems (Meade 1981, Morton 1968, Candy 1985).

1. Care should be taken in the choice of photomultiplier. The anode pulses should be as narrow as possible to allow high count rates. The pulse width is related to the focusing properties of the dynode structure as discussed in Chapter 2. Linear focused tubes are the best commonly available, fast, large cathode area photomultiplier tubes. It is preferable to have as high gain a tube as possible rather than using low gain tubes with external amplification in order to take advantage of the extremely fast amplification properties of the dynode structure. The quantum efficiency of the photomultiplier should be as high as possible, while the dark current should be as low as possible. This latter factor depends on the effective photocathode area and as small as possible a cathode area, consistent with experimental conditions, is preferable. Many manufacturers will now provide so-called photon counting photomultipliers and will even, for a small price, specially select very low dark-noise photon counting tubes for use in low level counting applications.

2. The design of the optical system should merit some attention. It should maximize the coupling between the light and the photomultiplier and should allow the use of small area photocathodes, to take advantage of their low dark noise properties. It is often desirable to image the aperture of the system onto the photocathode in order to prevent movement of the light over a possible nonuniform photocathode should there be relative motion between the object and the photomultiplier.

3. Radio frequency interference, both airborne and mains borne, is a major problem in photon counting systems, causing spurious pulses which are counted by the system. For this reason, the photomultiplier should be housed in an r.f. screened enclosure and the mains supply should be r.f. filtered. The connection between the photomultiplier and the amplifier should be a length of coaxial cable as short as possible in order to minimize pickup. If this is not possible, a preamplifier is usually mounted directly on, or as close as possible to, the photomultiplier. Correct impedance matching of cables and amplifiers should be employed to prevent reflections which would result in spurious multiple counts.

4. The high voltage power supply to the photomultiplier should be as low noise as possible. Some photomultiplier dynode chains have zener diodes between the photocathode and the first dynode and this can generate noise to very high frequencies. If this is the case, then it will be necessary to decouple this diode by means of a capacitor placed across it.

Ratio counting

In the direct counting mode, the counting time is set by an internal clock/present counter system. This mode cannot compensate for variations in the intensity of the source. One way to do this is to make the counting time depend on the light intensity. A second photomultiplier/discriminator system is used to monitor the light intensity and the measurement time is inversely proportional to the output of this photomultiplier.

Synchronous detection (Arechi et al. 1966)

In order to separate signal counts from background counts the signal to the photomultiplier can be modulated, for example by an optical chopper. During the period when the light is on the photomultiplier, the counts corresponding to signal plus noise can be sent to one counter; during the dark period, the counts corresponding only to noise can be sent to a second counter (Fig. 5.7). Subtracting the two counts should lead to a significant improvement in signal-to-noise ratio. As before, let the measurement time be T and suppose that this period is divided into a time kT, when the combined signal plus noise is measured, and $(1-k)T$ when only noise is recorded. There are n_s signal counts and n_B background counts during these intervals respectively. If there are $n_b(1-k)$ counts in the time $(1-k)T$, then, in the interval kT, there are $(k/1-k)n_b$ counts. Thus, if the total counts registered during the period in which the light is on the photo-multiplier is n_{TOT}, then the number of signal counts is

$$n_s = n_{TOT} - \left(\frac{k}{1-k}\right)n_B$$

If we assume Poisson statistics for signal and background, the error in signal counts in the period T is the sum of the variances of the two

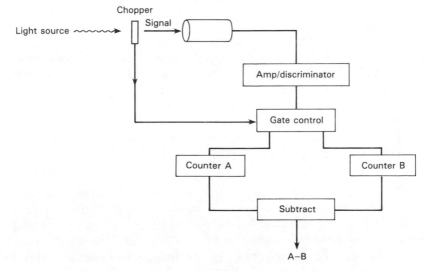

Fig. 5.7

contributions

$$\langle \Delta n_s^2 \rangle \langle n_{TOT} \rangle + \left(\frac{k}{1-k} \right)^2 \langle n_B \rangle$$

With a signal and background count rate of N_s and N_B counts per second, the signal-to-noise ratio is

$$\frac{S}{N} = \frac{N_s k T}{\left[(N_s + N_B)kT + \left(\frac{k}{1-k} \right)^2 N_B(1-k)T \right]^{1/2}}$$

$$= \frac{N_s k T}{\left[(N_s + N_B)kT + \frac{k^2}{1-k} N_B T \right]^{1/2}}$$

$$= \left\{ \frac{N_s k T}{\left(1 + \frac{N_B}{N_s} \frac{1}{1-k} \right)} \right\}^{1/2}$$

We may distinguish two limits:
(a) $N_s \gg N_B$. Then

$$\frac{S}{N} \approx (N_s k T)^{1/2}$$

which is maximized if k is made close to 1.
(b) $N_B \gg N_s$. Here

$$\frac{S}{N} \approx \left[\frac{N_s^2 k T (1-k)}{N_B} \right]^{1/2}$$

This is maximized when $k = \frac{1}{2}$. Thus, under these adverse conditions, a 50 percent duty cycle on the optical modulator should be used for maximum signal to noise, an ideal experimental arrangement for operation with mechanical or piezoelectric choppers.

5.1.2 Photon counting with photodiodes

Photon counting techniques are mainly used with photomultipliers, but below photon energies of about 1 eV photomultipliers are not readily available. There has, however, been some activity in the use of avalanche photodiodes to extend the range of photon energies over which photon counting can be used. In particular, germanium APDs have to be utilized for photon counting between 0.7 eV and 1.1 eV (see, for example, Fichtner and Hacker 1976, Haecker, Groezinger and Pilkuhn 1971). Avalanche multiplication for photocurrent can be regarded as the solid state analog of photomultipliers.

When used in the photon counting mode, the APDs are biased slightly *above* breakdown (known as the Geiger mode). In this mode, the photodiodes are much less sensitive to voltage instabilities than they are in their normal (sometimes called analog) mode of operation when they are biased *below* breakdown. In the Geiger mode, single photogenerated carriers trigger so-called microplasma avalanches. The resultant multiplication is so large that small variations in the voltage (and hence in the gain) become negligible. The avalanche will continue until space charge effects reduce the voltage to less than the breakdown voltage.

There are drawbacks to the use of APDs in the photon counting mode. At room temperatures, as the gain of the photodiode increases, the noise due to thermal electron multiplication increases faster than the signal. Cooling the photodiode would obviously help here, but many avalanche photodiodes fail to work properly at low temperatures. Some silicon avalanche photodiodes have been designed to work at liquid nitrogen temperatures, 77 K, and have a higher detected quantum efficiency than photomultipliers between 400 and 900 nm with marked superiority at longer wavelengths, although they are limited to count rates of about 10^4 s^{-1} in the photon counting mode (Ingerson *et al.* 1983). Another problem arises from trapping effects within the diode structure. Photoelectrons can be trapped at impurities or dislocations in the photodiode. These are liberated from the traps spontaneously with widely varying lifetimes and can then generate avalanches, leading to false counts.

The need in optical communication for detectors in the 1.3–1.6 μm wavelength range has led to the search for a solid state photomultiplier using materials other than silicon. The III-V semiconductors would be useful in this respect. However, the ionization rates for electrons and holes in these materials are nearly equal and consequently the noise level is very high. Recently, the novel superlattice materials have allowed the ionization rate of electrons and holes to be artificially altered. These devices consist of a stack of a large number of alternating layers of AlGaAs and GaAs sandwiched between a p-doped GaAs layer where electron–hole pairs are originally produced and an n-doped GaAs layer where the multiplied electrons are collected. At the interface between the AlGaAs and GaAs layers there are band discontinuities, these discontinuities being much greater for electrons than for holes. When an electron travels from an AlGaAs layer where there is negligible ionization to a GaAs layer, it sees a drop in the conduction band edge, causing a large reduction in electron ionization energy. The reduction in hole ionization energy is much smaller, so that the superlattice multiplies electrons more efficiently than holes, thus reducing noise. Such superlattice avalanche photodiodes seem to have an important future in photon counting (Capasso 1982, Chemla 1985).

5.1.3 Photon counting with image arrays

Photon counting can also be effected with image arrays. As before, enough pre-readout gain must be provided so that each photon event will produce an electron pulse which is significantly above the noise level so that discrimination can be employed. However, new problems arise with arrays. We must ensure that, at the maximum photon arrival rate, no more than one photon event will normally be registered on any picture element in a frame time and, in practice, this means that the frame scan rate must be many times the average rate of photon arrival. This restricts the technique to the detection of very faint images. A rapidly scanned image device with a large number of pixels means the system must have large signal video bandwidth, producing more readout noise per sample. This requires greater pre-readout gain in the imaging device for the requirements of noise discrimination. Finally, some form of computer needs to be used to record photon events by location in the image.

An example of this technique was that of Sandel and Broadfoot (1976) to produce a photon counting system based on such a two-dimensional photodiode array. The array used was a Reticon Corporation two-dimensional array possessing 1024 elements. The photodiodes were located on 100 μm centers and had a charge storage capacity of $\sim 5 \times 10^6$ electrons. The shift register could be clocked at rates between 10 kHz and 10 MHz. A photocathode/microchannel plate system was used as an image intensifier. The photoelectrons from the microchannel plate were then incident on a phosphor deposited on a fiber optics window. The photons produced were thus guided down the fiber system onto the photodiode array (Fig. 5.8).

In the photon counting mode, the sampling rate of each photodiode is determined by the decay lifetime of the phosphor (if a single event is not to be recorded in successive scans). For the phosphor used, a maximum clock frequency of 1.6 MHz was set. The whole system was packaged into a disk of diameter 5.1 cm and of thickness 2.6 cm. Computer control was

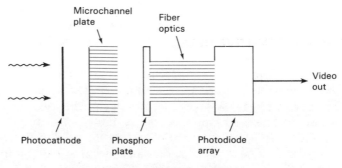

Fig. 5.8

used for clocking and storage of the video signal, and fixed pattern noise was subtracted in real time by a computer. Photon counting imaging systems have also been effected with vidicon systems but present developments in this area are to produce single photon counting imaging arrays using CCD technology. Progress in this area has mainly been fueled by the needs of astronomers, and some examples of practical systems are given in the references. In particular, it would appear that CCDs have a significant advantage over the other technologies since lower amplifier noise permits photon counting on reasonably sized arrays without the use of auxiliary image intensifiers, such as the microchannel plate (Boksenberg 1982).

If we refer again to Fig. 5.2, we can see that, at higher light levels, or with slower electronics, we are no longer able to count the individual photon pulses. In this case, in the presence of light, the output from the transducer is a d.c. signal with superimposed noise. This signal can be detected in several ways.

5.2 D.C. MEASUREMENT TECHNIQUE

This is perhaps the simplest available technique. For example, the signal current from a photomultiplier may be converted to a voltage as discussed in Chapter 4. The resultant noisy voltage may be further amplified and the signal-to-noise of the system improved by passing the signal through a simple low pass filter whose bandwidth (or time constant) can be adjusted by the operator. Similarly, the voltage output of a detector can be amplified and filtered.

Such techniques are widely available and used, but are susceptible to $1/f$ noise, drift, and dark current. There is no discrimination against these noise sources and the technique is usually limited to relatively high light levels. If the signal-to-noise ratio is so low that time constants of the order of ten seconds are needed, then we need to resort to other methods. The phase sensitive detector is one such method and this is discussed in the next section.

5.3 THE PHASE SENSITIVE DETECTOR

The phase sensitive detector (PSD) is an instrument used to detect and measure the amplitude and phase of a periodic signal. It can be used in experiments to enhance poor signal-to-noise ratios where a 'clean' reference voltage is available at the same frequency as the signal (Meade 1982, Blair and Sydenham 1975).

In order to understand the principle of operation of the PSD, consider Fig. 5.9. $V(t)$ is a sinusoidal voltage at some frequency ω; the switch is

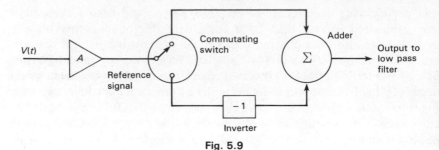

Fig. 5.9

switched by a reference voltage at the same frequency. One half of the signal per half cycle passes through an inverter and the two halves of the signal are recombined in the adder to give a resultant output. The output signal as a function of the phase ϕ between signal and reference are shown in Fig. 5.10.

From these diagrams, it can be seen that the resultant signal will give a d.c. voltage output when passed through a low pass filter which depends on ϕ. Mathematically, the switching is equivalent to multiplying the signal voltage by a square wave of equal positive and negative amplitudes. If the signal voltage is

$$V_s = V_0 \cos(t + \phi)$$

then the output voltage is of the form

$$V_{\text{out}} = V_0 \cos(\omega t + \phi) \times (\cos \omega t - \tfrac{1}{3} \cos 3\omega t + \tfrac{1}{5} \cos 5\,\omega t \ldots)$$
$$= V_0 \cos \phi + \text{terms of order } 2\omega \text{ and higher frequencies}$$

If this signal is passed through a low pass filter so that components of frequencies 2ω and higher are filtered out, then the output voltage is d.c.

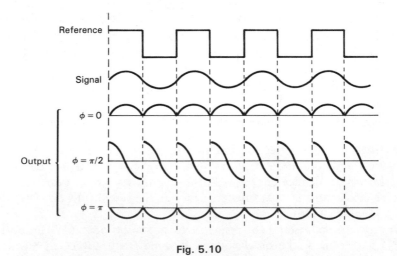

Fig. 5.10

of magnitude

$$V_{out} = V_0 \cos \phi$$

This is the basic equation of all phase sensitive detectors based on analog signals. We can envisage two possible situations, assuming the amplitude of the reference remains constant. First, if the phase is kept constant, the detector produces an output which is proportional to the signal amplitude. This is the use of the phase sensitive detector for measuring small amplitude signals, for example, in spectroscopy. Second, if the signal and reference amplitudes are kept constant, the output will be related to the phase difference ϕ between the reference and the signal. This mode of operation is used to relate spatial position in, for example, fringe positioning in interferometers.

If the signal and reference have *different* frequencies, then, applying the analysis above, the output voltage is of the form

$$V_{out} = V_0 \cos \ [\,(\omega_s - \omega_r)t + \phi\,) + \text{terms of higher frequency}$$

where ω_s = signal frequency and ω_r = reference frequency.

Note that this contains no d.c. component – only a.c. components. This component can be filtered out using a low pass filter; the closer are ω_s and ω_r, the narrower the bandwidth of the filter needed to filter out the component.

Thus, we can see how the PSD works. A signal with noise is modulated at some frequency and fed into the PSD. The modulated signal produces a d.c. output after passing through a low pass filter; the noise, however, produces only an a.c. voltage and this can be considerably attenuated on passing through the filter. Thus, the signal-to-noise ratio of a voltage can be improved by passing it through the PSD.

There is one further complication when the above switching is used in the PSD. If the signal frequency contains an odd multiple of the reference frequency then a d.c. output will be obtained from the PSD. For example, suppose $\omega_s = 3\omega_r$,

$$V_{out} = V_0 \cos \ (3\omega_r t + \phi) \times (\cos \ \omega_r t + \tfrac{1}{5}\cos \ 5\omega_r t + \ldots)$$

$$= \frac{V_0}{3} \cos \phi + \text{frequency-dependent terms}$$

(The frequency response of the PSD with square wave switching is shown in Fig. 5.11.)

Because of this, it is important not to modulate the signal such that its odd harmonics coincide with a strong interference component such as the mains frequency. It is possible to construct a PSD which does not use signal switching as above, but which electronically multiplies the modulated signal by a sinusoidal reference. This will not produce the problem with the harmonics; however, switching is more common since digital gating is easier to design and use for this purpose than multiplication by a sine wave.

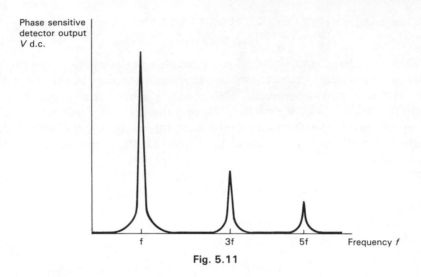

Fig. 5.11

Applications do exist, however, where it is not possible to use square waves, since the signal may not respond fast enough to be square wave modulated with adequate squareness.

A practical experimental signal processing system using a PSD is shown in Fig. 5.12. An optical signal is periodically modulated ('chopped') by a rotating slotted disk ('chopper'), and a reference signal is derived using a photoelectronic module consisting of an LED and phototransistor. The signal and reference can then be passed into the PSD.

It is important to recognize that the signal and reference are derived from the same source. The PSD is in effect a very narrow band filter whose bandwidth is dependent on the bandwidth of the output low pass filter.

This has a half-power bandwidth of

$$\Delta\omega = \frac{1}{\tau}$$

where $\tau = RC$ is the time constant of the filter. For example, with a time constant of 10 s, we obtain a half-power bandwidth of 0.032 Hz. Thus, if the modulating frequency is 10^3 Hz, then the output of the PSD is 3 dB down at 999.984 Hz and 1000.016 Hz. This response is equivalent to a filter

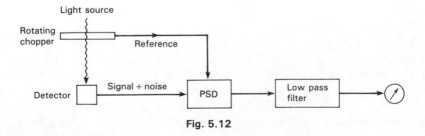

Fig. 5.12

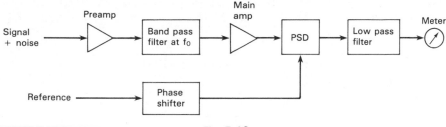

Fig. 5.13

with a Q value of $\sim 10^5$. Of course, we cannot increase the time constant of the filter indefinitely because the system response becomes extremely slow, as explained in the chapter on filters. This gives us a very highly tuned filter without the need for expensive components. However, because the reference voltage is derived from the signal, the centre of the PSD response is maintained at the signal frequency, i.e. we have 'locked-in' to the signal.

Also, since the PSD is an a.c. system, it will not suffer from drift problems associated with a d.c. system, and, by chopping at high enough frequencies, $1/f$ noise problems can be eliminated.

The practical system shown above is primitive, and a more complete system will now be discussed (see Fig. 5.13).

Most commercial systems have phase shifters for shifting the phase of the reference with respect to the signal to allow the experimenter to look for components of signal at reference frequency but with different phases. To get the reference and signal in phase; the phase shifter is varied until maximum signal is obtained and at this point the phase shifted reference and signal are in phase. In practice, most PSDs will have a button which can be depressed to add $90°$ to the phase of the reference. The phase shifter is varied until a null signal is obtained; the $90°$ button is then depressed to obtain in-phase signal and reference.

The band pass filter limits the size of signal and noise entering the main amplifier. This is important since, if the amplifier is driven into a nonlinear response, noise at different frequencies will be mixed together at the signal frequency, resulting in more noise.

The above use of the PSD requires a reference of correct (and known) phase; sometimes, a reference is available but its phase is uncertain or is varying such that a normal PSD would require constant adjustment and there would be uncertainty as to whether an output variation were a signal change or a phase change. In this case, it would be necessary to use a double PSD system as illustrated in Fig. 5.14.

The two PSDs are set in quadrature (i.e. $90°$ out of phase with each other) but both are triggered by the same reference. Thus, if the output of $(PSD)_1$ is $V_s \cos \phi$ then that of $(PSD)_2$ is $V_s \sin \phi$. The two outputs are then squared electronically and added. This gives an output signal

$$V_{out}^2 = V_s^2 \cos^2 \phi + V_s^2 \sin^2 \phi = V_s^2$$

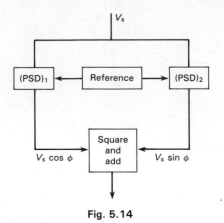

Fig. 5.14

5.4 REPETITIVE SIGNALS

The previous two techniques are essentially used when the light source is constant. However, it often happens that the light source is a repetitive signal; for example, if we were measuring the fluorescent decay of a solid which is excited by short duration repetitive light pulses. We may be interested in the shape or duration of the optical signal as well as its amplitude. Techniques to provide such information will now be discussed.

5.4.1 The 'boxcar' integrator

The 'boxcar' integrator, or single channel averager, is an instrument used to retrieve the waveforms of a repetitive signal from noise or to measure the amplitude of a repetitive pulse buried in noise. It can be used in two modes of operation – a scanning mode for use in waveform retrieval and in a single-point mode for pulse measurement. The boxcar samples the waveform once every cycle and the sample information is averaged so that the output due to the noise averages to zero. The basic arrangement is shown in Fig. 5.15 (Klauminzer 1975).

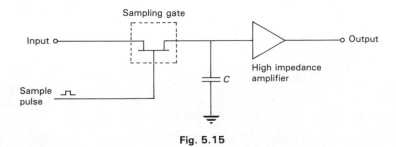

Fig. 5.15

The sampling gate can be switched into a low impedance state from the high impedance state by the application of a sampling pulse with sampling period T which is coherent with the signal. When the gate goes into the high impedance state (the sampling mode), the output signal will follow the input. The capacitor C charges up with time constant rC, where r is the resistance of the gate and signal source. When the gate goes into the high impedance state (the hold mode), the capacitor holds the output until the next sample is taken. In practice, this signal will leak away due to the finite conductance of the capacitor and the input bias requirement of the output amplifier, which is supplied by the capacitor during the hold period. The ratio of the useful hold time to sample time is proportional to R_{IN}/r, where R_{IN} is the input resistance of the amplifier. Typical values are $R_{IN} = 100$ MΩ and $r = 100\ \Omega$, giving $R_{IN}/r = 10^6$.

There are three operating modes for the boxcar, depending on the ratio of $R_{IN}C$ and the gatewidth T_s of the sampling pulses. The first case to consider is that of $rC \gg T_s$, as shown in Fig. 5.16.

If, at some time, the voltage on the capacitor is at V_1 and the sampling switch is closed, after a further time T_s, the voltage on the capacitor will have risen to V_2, where

$$V_2 = V_i(1 - e^{-T_s/\tau_1}) + V_1\,e^{-T_s/\tau_1}$$

Here V_i is the input voltage which is being sampled and $\tau_1 = rC$. On opening the sampling switch, V_2 will decay to a value V_3, where

$$V_3 = V_2\,e^{-(T-T_s)/\tau_2}$$

where $\tau_2 = R_{IN}C$.

If we ignore this leakage and assume that at $t = 0$ the voltage across C is zero, then after a time t when n samples have been taken using the assumption that $T_s \ll \tau_1$, the voltage on the capacitor is

$$V_C \approx nV_i\,\frac{T_s}{\tau_1}$$

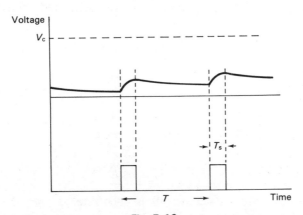

Fig. 5.16

The boxcar is now acting as a true integrator but with an effective time constant of

$$\tau_{\text{eff}} = rC\left(\frac{T}{T_s}\right)$$

The ratio T/T_s is sometimes known as the duty cycle. At the end of a suitable averaging period the output of the integrator is read and the capacitor is discharged.

The more common operating mode is the exponential weighted averaging, where $rC \geqslant T_s$. In this case, the voltage on the capacitor after n samples is

$$V_C = V_i(1 - e^{-nT_s/\tau_1})$$

with $\tau_1 \approx T_s$.

In this scheme, there is a weighting which favors the later samples. The convenience of this mode is that the output varies at a reasonable rate as the input changes, enabling a direct readout of the boxcar to be made.

The final mode of operating is that in which $rC \ll T_s$. In this case, the capacitor will charge up to the input voltage almost immediately. This mode is sometimes known as the sample-and-hold mode. The value obtained at each sample can be stored in some form of memory e.g. using a digital computer) and appropriate averages taken (see Chapter 6).

The technique by which boxcars improve the signal-to-noise ratio is known as signal averaging. In all the modes of boxcar operation, samples of a waveform which will contain both signal and noise are added together, with an average being taken to give the final value. It will now be shown that, if n samples are taken and averaged, there will be an enhancement of the signal-to-noise ratio by $\sqrt{n}$.

Let us suppose that we have a repetitive signal with associated noise which we will assume is Gaussian. We sample the signal at the same point in each cycle and that the value of the signal at that point is S. Superimposed on this signal is some noise. Thus, the measured value on the kth sample is

$$F(t_k) = S + N(t_k)$$

$N(t_k)$ has a mean value of zero and an r.m.s. value of σ, the standard deviation. After taking m samples, the average measured value is

$$\frac{1}{m}\sum_{k=1}^{m} f(t_k) = \frac{1}{m}\sum_{k=1}^{m} S + \frac{1}{m}\sum_{k=1}^{m} N(t_k)$$

$$= S + \frac{1}{m}\sum_{k=1}^{m} N(t_k)$$

Assuming the noise in each sample is uncorrelated, then the total noise from k samples is obtained by adding the mean square of the noise and taking the square root:

$$\sum_{k=1}^{m} N(t_k) = \sqrt{m\sigma^2}$$

$$\therefore \quad \frac{1}{m} \sum_{k=1}^{m} f(t_k) = S + \frac{\sigma}{\sqrt{m}}$$

Thus, after m samples, the signal-to-noise ratio is

$$\frac{S}{N} = \sqrt{m} \, \frac{S}{\sigma}$$

Now, S/σ is the signal-to-noise ratio for one sample; thus, by repetitively sampling once each cycle for m cycles, we have increased the signal-to-noise by a factor of $\sqrt{m}$.

The above theory essentially describes the single-point mode of the box-car and will enable the measurement of the mean amplitude of repetitive pulses buried in noise. An example of such an experimental setup to measure the pulse intensity of a repetitive train of light pulses is shown in Fig. 5.17. The photodiode produces a coherent trigger which can be delayed manually and its width varied to coincide with and enclose the light pulse.

In this mode, the boxcar is very similar to the phase sensitive detector except that the phase sensitive detector is on for 50 percent of the cycle. If the pulse signal may only be on for a small fraction of the repetition period, then the phase sensitive detector would spend most of its time measuring noise only. Thus, in this case, the boxcar must represent a significant im-provement in signal-to-noise over the phase sensitive detector.

In the boxcar scan mode, the waveform is sampled and averaged at each point on its waveform in turn, moving to the next point when the output has reached the voltage of the waveform at that point with sufficient signal-to-noise. The added modification to the single-point mode system is a unit which changes the delay of the gate at a set rate. These are realized on com-mercial instruments via the timebase, which can be set to give a full scan of the gate across one cycle of the waveform, and the scan control, which fixes the time taken for a complete scan of the gate across one cycle of the

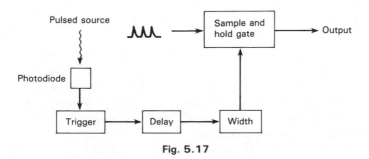

Fig. 5.17

waveform. (An initial delay is also usually available to allow the operator to set the starting point of the scan.) If the waveform has fine structure, the gate width should, of course, be smaller than the width of this structure. The width of this pulse sets the highest frequency components in the waveform which can be measured by the boxcar. The gate time constant rC can also usually be set to one of several values; a larger time constant means the voltage fluctuations on the storage capacitor due to noise inputs will be decreased, although the measurement with take longer.

It would seem from the preceding discussions that the phase sensitive detector is simply a special case of the boxcar having a 50 percent duty cycle. However, there is a subtle difference between phase sensitive techniques and signal averaging. The phase sensitive detector acts by restricting the bandwidth of the system to a value centered about the reference frequency, relying on the fact that the noise is 'white' so that the measured noise is proportional to the bandwidth of the detector. It is used with narrow band signals and will be ineffective where the signal is complex and broad band, or where the noise contains strong components which overlap the signal frequency components. Such restrictions do not apply to signal averagers, which can provide enhancement or the signal-to-noise ratio of wide bandwidth signals even in a frequency range containing strong noise components. In frequency space, the response function of the signal averager is that of a comb filter with its teeth centered at the frequency components of the signal and with a passband which is inversely proportional to the sampling period and the number of samples (Parisi 1985).

5.4.2 The multichannel analyzer

The multichannel analyzer recovers a repetitive signal from noise by dividing each signal cycle into a number of equal segments, the number of segments determined by the resolution required. It will consist of n sample-and-hold gates where n is the number of segments. On each cycle, each storage capacitor charges towards the mean voltage of the appropriate segment. The gates must be opened by sequential logic triggered by the coherent reference pulse.

The boxcar in scan mode will perform the same function as the multichannel analyzer. However, since the boxcar only samples at one point and the scan is continuous, the boxcar ignores a large part of the waveform between sampling pulses. The boxcar will take n times as long to plot out a given waveform with the same signal-to-noise as an n-channel multichannel analyzer. However, the boxcar is considerably cheaper than the multichannel analyzer. The complexity of the multichannel analyzer also means the sampling gates need to be fairly simple and will not have the fast sampling capability of the boxcar.

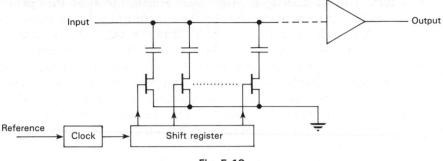

Fig. 5.18

5.5 OPTICAL HETERODYNE TECHNIQUES
(Teich 1968, Smith 1985)

No matter how much signal processing we perform, there is always a fundamental level of noise below which we cannot go. In photomultipliers, for example, this is due to shot noise associated with photocurrent emitted from the photocathode, and this will limit the smallest optical signal we can measure. This is known as the *quantum limit of optical detection*. The signal processing techniques discussed so far will not approach this limit since, as we have seen in Chapter 2, this would imply the total suppression of dark current and background radiation.

The optical heterodyne technique is one which will enable us to achieve this quantum limit. The experimental arrangement employed is shown schematically in Fig. 5.19.

Two optical signals at frequencies ω_1 and ω_2 fall on the photodetector whose nonlinear properties mix the two waves to produce a signal at the difference frequency $\omega_1 - \omega_2$. If the beam at ω_2 (usually produced locally with a laser beam and so known as the local oscillator) is very much stronger than the signal term at ω_1, then the shot noise due to the local oscillator can dwarf all other terms, thus achieving the quantum limit of detection. In the

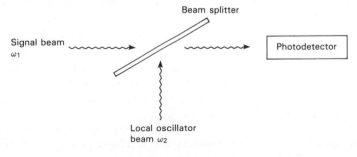

Fig. 5.19

following sections we shall obtain expressions for the minimum detectable power using the photomultiplier and the photoconductor in the heterodyne technique.

5.5.1 Heterodyne detection in photomultipliers

Consider the two electromagnetic waves at frequencies ω_1 and ω_2 impinging on a photomultiplier cathode. The total electic field at the the photocathode is

$$E_t = E_1 \cos \omega_1 t + E_2 \cos \omega_2 t$$

where E_1 and E_2 are the amplitudes of the individual incident waves. Assuming that these two waves have the same polarization, the photo-current produced in the detector is proportional to the square of the total electric field

$$I_{ph} = kE_t^2 = k[E_1^2 \cos^2 \omega_1 t + E_2^2 \cos^2 \omega_2 t$$
$$+ E_1 E_2 \cos(\omega_1 + \omega_2)t + E_1 E_2 \cos(\omega_1 - \omega_2)t]$$

where k is a constant.

The detector cannot follow the instantaneous intensity at optical frequencies and so will average the terms in ω_1, ω_2, and $\omega_1 + \omega_2$. However, we shall assume that the photomultiplier has a sufficiently high frequency response to follow the signal at the difference frequency, i.e.

$$I_{ph} = k\left[\frac{E_1^2}{2} + \frac{E_2^2}{2} + E_1 E_2 \cos(\omega_1 - \omega_2)t\right]$$

If we further assume that the intensity of the local oscillator is very much greater than that of the signal, i.e. that $E_1 \ll E_2$, then

$$I_{ph} = k\frac{E_2^2}{2} [1 + 2(E_1/E_2)\cos(\omega_1 - \omega_2)t]$$

$$= k\frac{P_2}{2} [1 + 2\sqrt{(P_1/P_2)}\cos(\omega_1 - \omega_2)t]$$

in terms of the optical power P. We can now evaluate the constant k by setting $P_1 = 0$. In this case, the photocurrent will be

$$I_{ph} = (\eta e P_2/h\omega_2) = kP_2/2$$

so that

$$k = (2\eta e/h\omega_2)$$

The photocurrent at the photomultiplier in the presence of both optical fields consists of a d.c. term and a term which oscillates sinusoidally at the difference frequency $\omega_1 - \omega_2$. This frequency is known as the *heterodyne frequency.* If we now confine measurement of the signal to a pass band about the heterodyne frequency, the mean square photocurrent 'signal' is

$$\langle I_{ph}^2 \rangle = 0.5(\eta e/h\omega_2)^2 . 4P_1P_2$$
$$= 2(\eta e/h\omega_2)^2 P_1 P_2$$

The photocathode shot noise consists of two terms, one due to the cathode dark current, the other due to photon noise

$$\langle i_N^2 \rangle = 2e(\langle I_d \rangle + \langle I \rangle)\, \Delta f$$

The mean cathode photocurrent is given by

$$\langle I \rangle = (\eta e P_2/h\omega_2)$$

so that the signal-to-noise power ratio at the anode of the photomultiplier is

$$S/N = \frac{2G^2 P_1 P_2 (e\eta/h\omega_2)^2}{\{G^2 2e(\langle I_d \rangle + \eta e P_2/h\omega_2) + 4kT/R_L\}\, \Delta f}$$

where R_L is the anode load resistor. If we now make the local oscillator power P_2 so large that the shot noise due to it dominates the noise, then

$$S/N \approx \frac{2G^2 P_1 P_2 (e\eta/h\omega_2)^2}{G^2 2e(\eta e/h\omega_2)P_2\, \Delta f} = \frac{P_1 \eta}{h\omega_1\, \Delta f}$$

where we have used the approximation $\omega_1 \approx \omega_2$. The minimum detectable power is therefore

$$(P_1)_{min} = (h\omega_1\, \Delta f/\eta)$$

A similar expression will hold for reverse-biased photodiodes where the noise is of the same form as encountered with photomultipliers.

5.5.2 Heterodyne detection in photoconductors

We can obtain a similar expression for the minimum detectable power using photoconductors in the heterodyne mode. However, for a photoconductor, the noise behavior differs from simple shot noise and the above results are not directly applicable. We shall assume that generation–recombination noise is the dominant source of noise in this case.

To obtain an expression for the photocurrent generated in the photoconductor, we must solve the rate equation for the generation of photocarriers

$$\frac{dN_c}{dt} = kE_t^2 - \frac{N_c}{\tau_0}$$

where

$$E_t = E_i \cos \omega_1 t + E_2 \cos \omega_2 t$$

As before, the photoconductor averages the optical frequency terms so that

$$\frac{dN_c}{dt} = k\left[\frac{E_1^2}{2} + \frac{E_2^2}{2} + E_1 E_2 \cos(\omega_1 - \omega_2)t\right] - \frac{N_c}{\tau_0}$$

$$\approx k\frac{P_2}{2}\left[1 + 2\sqrt{(P_1/P_2)}\cos(\omega_1 - \omega_2)t\right] - \frac{N_c}{\tau_0}$$

for strong local oscillator power. The photocurrent is given by

$$I_{ph} = N_c\frac{e}{\tau_d}$$

$$= \frac{e\eta}{\hbar\omega_2} \cdot \frac{\tau_0}{\tau_d}\, P_2\left[1 + 2\sqrt{(P_1/P_2)} \cdot \frac{\cos(\omega t - \phi)}{(1 + \omega^2\tau_0^2)^{1/2}}\right]$$

where

$$\tan\phi = \omega\tau_0$$

and we have written, for convenience, $\omega = \omega_1 - \omega_2$.

The mean square signal term is

$$\langle I_{ph}^2\rangle = \frac{1}{2} \cdot \frac{4P_1 P_2}{(1 + \omega^2\tau_0^2)}\left(\frac{e\eta\tau_0}{\hbar\omega_2\tau_d}\right)^2$$

and the generation–recombination noise is

$$\langle i_{GR}^2\rangle = \frac{4e^2\eta\left(\dfrac{\tau_0}{\tau_d}\right)^2(P_1 + P_2)\,\Delta f}{(1 + \omega^2\tau_0^2)}$$

The signal-to-noise power ratio is

$$\frac{S}{N} = \frac{2P_1 P_2}{(1 + \omega^2\tau_0^2)}\left(\frac{e\eta\tau_0}{\hbar\omega_2\tau_d}\right)^2 \cdot \frac{(1 + \omega^2\tau_0^2)}{4e^2\eta(\tau_0/\tau_d)^2(P_1 + P_2)\,\Delta f}$$

$$= P_1 P_2\eta/[2(P_1 + P_2)\hbar\omega_2\,\Delta f]$$

When $P_1 \ll P_2$

$$\frac{S}{N} \approx \frac{P_1\eta}{2\hbar\omega_1\,\Delta f}$$

giving a minimum detectable signal of

$$(P_1)_{min} = 2\hbar\omega_1\,\Delta f/\eta$$

Thus these devices are a factor of two less sensitive than a photoemitter or a reverse-biased photodiode of the same quantum efficiency.

For a photoconductor operating at a wavelength of $1\ \mu m$ and with $\Delta f = 1$ Hz and $\eta = 1$

$$(P_1)_{min} = 10^{-18}\ \text{W}$$

In any practical optical heterodyne system the theoretical performance

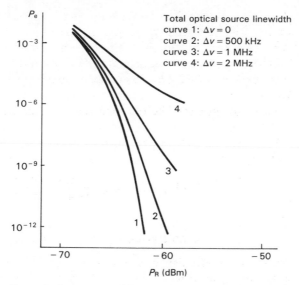

Fig. 5.20 (From G. Nicholson, 1984. *Electron Letters.* Vol. 20, pp. 1005–7)

will not be realized. The experimental arrangement requires careful alignment between the signal and local oscillator beams to maintain a constant phase over the surface of the photodetector. We should therefore expect that optical heterodyne detection will be relatively insensitive for thermal radiation. The polarization of the local oscillator should also be well matched to that of the input signal.

Heterodyne detection is currently being investigated for use in optical fibre communication systems. Several problems have to be overcome in such systems to achieve the optimum performance. Firstly, there may be insufficient local oscillator power to achieve shot noise limited detection from a receiver with high circuit noise. Shot noise limited detection can however be achieved using 1 μW of local oscillator signal with a low noise photodiode/preamplifier receiver combination. Secondly, the polarization of the signal on traversing the optical fibre will suffer fluctuations due to temperature variations and mechanical vibrations of the optical fiber. Thirdly, a more significant problem is that of the laser linewidth (i.e. signal plus local oscillator linewidth). For a given digital transmission rate (number of bits transmitted per second), the signal-to-noise ratio can be used to calculate the bit error rate (the precise expression depends on the modulation scheme employed). Figure 5.20 shows the dependence of the error rate on the receiver sensitivity for different laser linewidths for a bit rate of 140 Mbits per second. It shows that the optical receiver sensitivity depends strongly on the total optical linewidth for a given error rate. It is the improvement of semiconductor laser spectral qualities which is making optical heterodyne techniques increasingly important in optical communication systems.

SUMMARY

In this chapter we have been considering some aspects of signal processing. The signal quality has been improved using amplification and filtering, as described in Chapter 4, and this chapter was concerned with manipulation of the signal in some way. The advent of the microprocessor has revolutionized the area of signal processing and has made obsolete much instrumentation formerly used in signal processing. However, there are still some areas where systems not relying on microprocessors are still widely used. The approach adopted in this chapter has been to consider signal processing initially with very low light intensities where individual photons can be resolved by some detectors, notably the photomultiplier. In this case, the most efficient signal processing method is photon counting, which is essentially a digital technique. As the light intensity increases, the rate of photon arrival at the detector is so great that individual photon events can no longer be resolved and the output from the detector looks like a noisy, continuous one. Several approaches can then be adopted. The simplest one is to pass the signal through an appropriate filter. If the signal can be modulated in some way, phase sensitive detection can be employed. The phase sensitive detector acts as a high Q (quality factor) band pass filter centered around the signal frequency which rejects all noise lying outside the pass band of the filter. If the signal is a repetitive one, such as a pulsed light source, boxcar detection or multichannel averaging can be used. These techniques use the principle of signal averaging, whereby N averages of the signal are taken coherently with the signal, thus improving the signal-to-noise ratio by a factor $\sqrt{N}$. Although it is often stated that phase sensitive detection is a special case of the boxcar averager, the differences between the two techniques were emphasized. Finally, coherent techniques for the detection of light (the so-called optical heterodyne techniques) were discussed. These techniques are becoming increasingly important because of their use in optical communication systems.

REFERENCES

Amoss, J. and Davidson, F. 1972. Detection of weak optical images with photon counting techniques. *Applied Optics,* Vol. 11, pp. 1793–1800.

Arechi, F.T., Gatti, E., and Sona, A. 1966. Measurement of low light intensities by synchronous single photon counting. *Revue of Scientific Instruments*, Vol. 37, pp. 942–5.

Blair, D.P. and Sydenham, P.H. 1975. Phase sensitive detection as a means to recover signals buried in noise. *Journal of Physics E: Scientific Instruments*, Vol. 8, pp. 621–7.

Candy, B.H. 1985. Photomultiplier characteristics and practice relevant to photon counting. *Revue of Scientific Instruments*, Vol. 56, pp. 183–93.

Capasso, F. 1982. The chanelling avalanche photodiode: a novel ultra-low-noise interdigitated p-n junction detector. *IEEE Transactions on Electronic Devices ED-29*, pp. 1388–95.

Chemla, D.S. 1985. Quantum wells for photonics. *Physics Today,* May, pp. 56–64.

Fichtner, W. and Hacker, W. 1976. Time resolution of Ge avalanche photodiodes operating as photon counters in delayed coincidence. *Revue of Scientific Instruments,* Vol. 47, pp. 374–7.

Foord, R., Jones, R. and Oliver, C. 1969. The use of photomultiplier tubes for photon counting. *Applied Optics,* Vol. 8, pp. 1965–89.

Haecker, W., Groetzinger, O. and Pilkhun, M.H. 1971. Infrared photon counting by Ge avalanche photodiodes. *Applied Physics Letters.* Vol. 19, pp. 113–5.

Ingerson, T.E., Kearney, R.J. and Coulter, R.L. 1983. Photon counting with photodiodes. *Applied Optics,* Vol. 22, pp. 2013–8.

James, J.F. 1967. On the use of photomultiplier tubes as a photon counter. *Monthly Notices of the Royal Astronomical Society,* Vol. 137, pp. 15.–23.

Jones, R., Oliver, C.J. and Pike, E.R. 1971. Experimental and theoretical comparison of photon counting and current measurement of light intensity. *Applied Optics,* Vol. 10, pp. 1673–80.

Klauminzer, G.K. 1975. The basics of boxcars. *Laser Focus,* November, pp. 35–9.

Meade M.L. 1981. Instrumentation aspects of photon counting applied to photometry. *J. Phys. E: Sci. Inst.* Vol. 14, pp. 900–18.

Meade, M.L. 1982. Advances in lock-in amplifiers. *J. Phys. E: Sci. Inst.* Vol. 15, pp. 395–403.

Morton, G.A. 1986. Photon counting. *Applied Optics*, Vol. 7, pp. 1–10.

Oliver, C.J., and Pike, E.R. 1968. Measurement of low light flux by photon counting. J. Phys. D. Vol. 1, pp. 1459–68.

Parisi, J. 1985. High-performance waveform sampling analysis. *Rep. Prog. Phys.* Vol. 48, pp. 709–36.

Sandel, W.R. and Broadfoot, A.L. 1976. Photoelectron counting with an image intensifier tube and self scanned photodiode array. *Applied Optics,* Vol. 15, pp. 3111–4.

Smith, D.W. 1985. Coherent fibreoptic communications. *Laser Focus,* November, pp. 92–106.

Teich, M.C. 1968. Infrared heterodyne detection. *Proceeding of the IEEE,* Vol. 56, pp. 37–46.

PROBLEMS

5.1 A photoconductor at room temperature is to be used as a photon counter. It has a load resistance R which has a stray capacitance to earth of C. Assuming that the bandwidth of the system is determined by the time constant RC, and that the noise is dominated by Johnson noise in the load resistance, calculate the value of C to obtain a signal-to-noise ratio of unity if there is an arrival rate of one photon during a time interval of one RC time constant.

5.2 A photon counting system is to be used in the synchronous detection mode. The total time available for measurement is 1 ms. If the background count rate is 10^4 s^{-1} and the signal count rate is 10^3 s^{-1}, calculate the ratio of the interval of time for measurement of combined signal and background to that for background along to maximize the signal-to-noise ratio.

5.3 If the average number of photons arriving at a photodetector from a light source is 100 per second, how long must the photons be counted for in order to make a measurement accurate to 1 percent. (Assume Poisson statistics for photon emission.)

5.4   In a boxcar integrator, the 'on' resistance of the gate is 100 Ω and the input resistance of the buffer amplifier is 10^7 Ω. The width of the sampling pulse is 10 ns and the sampling period is 100 μs. A d.c. voltage is applied to the input of the boxcar. Obtain an expression for the mean output voltage of the boxcar and calculate its value if the sampling capacitor has a value of (a) 100 pF, (b) 10 nF.

5.5   A phase sensitive detector has a square wave reference and a square wave signal at the same frequency but at some phase ϕ with respect to the reference. Obtain an expression for the d.c. output of the phase sensitive detector.

6

Microprocessors and Signal Processing

6.1 INTRODUCTION

The development of microprocessors and microcomputers over the last several years has led to a revolution in signal processing. Microcomputers are used widely in laboratories to perform digital control and measurement. Complex digital circuits are low cost, accurate, and simple to implement. The manipulation of data as digital rather than analog signals allows great accuracy without problems of drift, linearity, component tolerances, etc., which are some of the many problems associated with analog circuitry. Because of the importance of such devices, we are going to look at the way in which we can convert the analog signal, obtained (suitably amplified and filtered) form our transducer, to digital form.

We should perhaps first distinguish between analog and digital systems. In an analog system, the actual value of, say, a voltage is important. In a digital system, that voltage would be present in binary form; that is, in one of two possible conditions. In electronic digital systems, these two states are given the values 0 and 1. The 0 and 1 are, in practice, represented by voltage ranges. In one of the most common form of digital circuits based on bipolar transistors (known as TTL), 0 would represent a voltage between 0 volts and 0.8 V while 1 would represent a voltage between 2 and 5 V. Thus, in our digital system 3 V and 4 V would represent the same state (1) but would be different in an analog system. In a digital system, the exact value of the voltage is not important, while in analog circuitry the precise value of the voltage carries significant information.

The above discussion shows some of the advantages of using digital systems. One of the greatest advantages is the simplicity of design compared with analog signals. Furthermore, digital systems are far more stable than analog systems. Highly complex circuits can be integrated into single integrated circuits, ranging from small scale integrated circuits, containing about ten transistors, to large scale integrated circuits, such as microprocessors, containing tens of thousands of transistors. These integrated circuits or 'chips' are built from a small number of digital building blocks, unlike analog integrated circuits, so that the degree of integration is far greater in digital chips. Of course, digital circuits suffer from a major problem; most signals encountered are analog in nature. In order to use them in a signal processing, we must convert the analog signal to a digital one (and vice versa, if we wish to read out the processed signal to some analog recording device). The interface between the analog and digital worlds are digital to analog converters (DACs) and analog to digital converters (ADCs) – collectively known as data converters.

6.2 SAMPLING, QUANTIZATION, AND CODING

In order to convert our analog signal to digital form, there are three basic steps. First, we must take a sample of the signal which is to be converted into digital form. Second, we must quantize that sample; that is, we must divide up the signal into a finite number of amplitude levels, corresponding to discrete values of the voltage from its minimum to maximum values. Finally, we must assign a code to a given voltage.

6.2.1 Sampling and Shannon's theorem

The first step in the conversion process is to take samples of the signal. It is legitimate to inquire whether information is lost in this process, and we would like to know the minimum sampling rate we can apply in order not to degrade the signal. This question is answered by the sampling theorem, first expressed by Shannon (1948).

Let us consider a continuous signal $f(t)$. We will sample this periodically with a sampling rate f'. If the time required to take each sample is τ, then

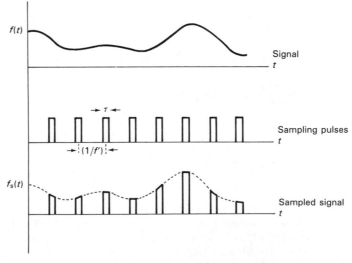

Fig. 6.1

the sampling process is mathematically equivalent to multiplying the signal by a series of unit pulses of width τ, as shown in Fig. 6.1, to produce a sampled signal $f_s(t)$.

We shall assume that the signal $f(t)$ is bandwidth limited to Δf, i.e. that it contains no signal components beyond a frequency Δf. The bandwidth limitation introduces no significant errors in the analysis, and, in practice, sharp cut-off low pass filters are introduced before the sampling process to ensure that the bandwidth limited condition is obeyed.

The sampled signal can be written

$$f_s = f(t)h(t)$$

where $h(t)$ represents the periodic sampling pulses

$$h(t) = \tau f' \left(1 + 2 \sum_{n=1}^{\infty} \frac{\sin n\pi\tau f'}{n\pi\tau f'} \cos 2\pi n f' t\right)$$

This frequency spectrum of the sampled signal can be obtained by taking the Fourier transform of $f_s(t)$, i.e.

$$F_s(\omega) = \int_{-\infty}^{+\infty} f_s(t)e^{-j\omega t}\,dt$$

$$= \int_{-\infty}^{+\infty} f(t)\tau f' \left(1 + 2 \sum_{n=1}^{\infty} \frac{\sin n\pi\tau f'}{n\pi\tau f'} \cos 2\pi n f' t\right)$$

$$= \int_{-\infty}^{+\infty} \tau f' f(t)e^{-j\omega t}\,dt$$

$$+ \sum_{n=1}^{\infty} \int_{-\infty}^{+\infty} 2\tau f' f(t) \frac{\sin n\pi\tau f'}{n\pi\tau f'} \cos 2\pi n f' t \, e^{-j\omega t}\,dt$$

Consider the integral of the mth term of the series:

$$\int_{-\infty}^{+\infty} 2\tau f' f(t) \frac{\sin m\pi\tau f'}{m\pi\tau f'} \cos 2\pi m f' t e^{-j\omega t}\,dt$$

$$= \tau f' \frac{\sin m\pi\tau f'}{m\pi\tau f'} \int f(t)e^{-j\omega t}(e^{j2\pi m f' t} + e^{-j2\pi m f' t})\,dt$$

$$= \tau f' \frac{\sin m\pi\tau f'}{m\pi\tau f'} \left\{\int_{-\infty}^{+\infty} f(t)e^{-j(\omega - 2\pi m f')t}\,dt \right.$$

$$\left. + \int_{-\infty}^{+\infty} f(t)e^{-j(\omega + 2\pi m f')t}\,dt\right\}$$

writing the cosine as the sum of two exponentials.

The Fourier transform of the sampled signal is thus

$$F_s(\omega) = \int_{-\infty}^{+\infty} \tau f' \, f(t) e^{-j\omega t} \, dt$$

$$+ \int_{-\infty}^{+\infty} \tau f' \sum_{\substack{n=-\infty \\ n \neq 0}}^{+\infty} \frac{\sin n\pi\tau f'}{n\pi\tau f'} \, f(t) e^{-j(\omega - 2\pi n f')t} \, dt$$

$$= \tau f' \, F(\omega) + \tau f' \sum_{\substack{n=-\infty \\ n \neq 0}}^{+\infty} \frac{\sin n\pi\tau f'}{n\pi\tau f'} \, F(\omega - n\omega')$$

where $\omega' = 2\pi f'$.

In frequency space, therefore, the sampled signal consists of the original signal together with all harmonics of the sampling frequency (Fig. 6.2).

From this diagram we can see that, as long as the different 'bands' are well separated, a suitable low pass filter can remove the harmonics, leaving the original signal. No information has thus been lost since the original signal can be reconstructed from the sampled data. In order that this may be so, then

$$f' - \Delta f > \Delta f$$

as is obvious from Fig. 6.2. This equation reduces to

$$f' > 2 \, \Delta f$$

which shows that the sampling frequency must be at least twice the maximum frequency present in the signal.

If the sampling frequency is not high enough, part of the spectrum centered about f' will overlap or fold over into the original signal. In the process of recovering the original signal, the folded part of the spectrum causes distortion of the recovered signal which cannot be eliminated. This effect is known as aliasing and the extra (distortion) frequency components introduced into the recovered signal are known as alias frequencies. Figure 6.3 shows how an inadequate sampling rate can introduce aliasing.

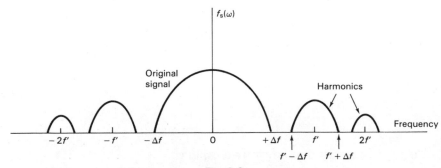

Fig. 6.2

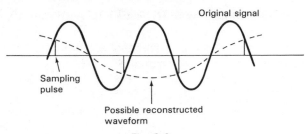

Fig. 6.3

Aliasing can be eliminated either by using a high enough sampling rate or by filtering the signal before sampling to limit its bandwidth to $f'/2$. In practice, there is always some aliasing present due to high frequency signal components, noise and nonideal, pre-sample filtering. The sampling rate is usually higher than that required by the sampling theorem but should not be significantly larger. If it is, no additional information will be obtained but more noise will be included than is necessary because the bandwidth has been increased by the higher sampling rate.

The sampling theorem can be extended to nonperiodic sampling. In this case, it can be shown that any $2\Delta f$ independent samples per second of a signal will completely characterize a bandwidth-limited signal. Alternatively, any $2\Delta f T'$ independent samples are needed to specify completely a signal over an interval T' seconds long.

6.2.2 Quantization

We must now divide up the voltage range so that we may relate the analog signal to a digital value. Conventionally, we divide the range into 2^n equal intervals. Typical values of n are 8, 10, 12, and 16.

Let us suppose that the minimum voltage in our signal is 0 V and the maximum is 10 V. Furthermore, we will take $n = 4$ for mathematical simplicity. Our converter is said to have four-bit resolution. We have divided our voltage range up into $2^4 = 16$ states. Each analog quantization state or quantum is

$$Q = \frac{\text{Full scale range}}{2^n} = \frac{10}{16} = 0.625 \text{ V}$$

6.2.3 Coding

The final step is to assign a digital code to each quantized level. The most popular code is natural binary, usually in its fractional binary form. In this case, we represent a number N as

$$N = a_1 2^{-1} + a_2 2^{-2} + \ldots + a_n 2^{-n}$$

where the coefficient a_j may be 0 or 1 and n is the resolution of the converter. This fractional scale is convenient for data converters, since the digital code can be interpreted as a fraction of full scale. Thus, a number in our four-bit converter may be written

$$0.1010$$

Conventionally, the point is dropped and the number written

$$1010$$

This number is

$$1 \times 2^{-1} + 0 \times 2^{-2} + 1 \times 2^{-3} + 0 \times 2^{-4} = 0.625$$

i.e. 62.5 percent of the full scale value.

The leftmost bit of the binary code is known as the most significant bit (MSB) while the rightmost bit is the least significant bit (LSB). The analog value of the LSB is

$$\text{LSB} = \frac{\text{Full scale range}}{2^n}$$

which is, of course, the same as the quantum Q. We should note that the maximum value of the digital code, when all the bits are at the value 1, does not correspond to the full analog scale but one LSB less than full scale. Thus, for our four-bit converter having a full scale of 10 V, the maximum analog value of the converter corresponding to the digital code 1111 is $(10 - 0.625)\ \text{V} = 9.375\ \text{V}$.

Several other coding schemes can be used with data converters. Binary coded decimal (BCD) is the code used where digital displays must be interfaced to ADCs, for example in digital voltmeters. In this scheme, four bits of the code are used to represent each decimal digit. Thus, only 10 out of the possible 16 codes are used. For example, the decimal number 329 is represented in BCD by

$$0011\ 0010\ 1001$$

The LSB for BCD is given by

$$\text{LSB} = \frac{\text{Full scale range}}{10^n}$$

where n is the number of decimal digits.

Such coding is usually used with an additional over-range bit; for example, a converter with a decimal full scale range of 999 with an over-range bit will have a full scale range of 1999 with a maximum output code of

$$1\ 1001\ 1001\ 1001$$

The additional bit doubles the range of the ADC and is commonly referred to as a $\frac{1}{2}$ digit. The resolution of this ADC is then $3\frac{1}{2}$ digits. The range can be further doubled by a second over-range bit to give a full scale of 3999. The converter will then have $3\frac{3}{4}$ digits resolution.

6.3 TRANSFER FUNCTIONS OF DATA CONVERTERS

In this section we will consider briefly the transfer function of an ideal data converter, i.e. the relationship between the input data and output data of the converter. For a DAC, there is a direct correspondence between the

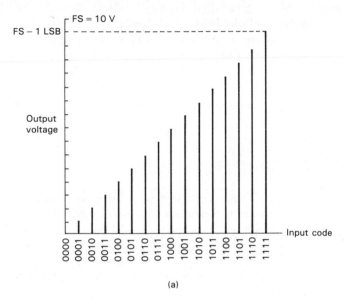

(a)

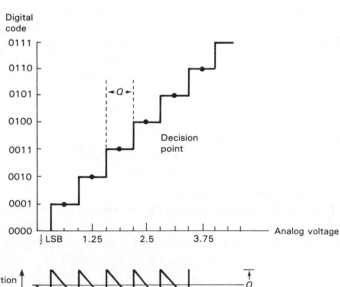

(b)

Fig. 6.4

digital code applied to the input and the analog output. For an ideal four-bit converter with a 10 V full scale, a digital input code of 1010 will produce an analog output voltage of 6.25 V. The 16 possible output voltages from the four-bit converter are shown in Fig. 6.4(a). The transfer function of a four-bit ADC is shown in Fig. 6.4(b). The one-to-one correspondence of the DAC does not apply. A given digital code corresponds to a range of analog voltages. Compared with the DAC, the transfer characteristic of the ADC is offset by $+\frac{1}{2}$ LSB along the analog axis. This ensures that the transitions between digital codes occur $\frac{1}{2}$ LSB on either side of the input for a particular code. Thus, for the four-bit converter, the 0000 to 0001 code transition occurs at 0.3125 V, the 0001 to 0010 code transition occurs at 0.9375 V, and so on. The nominal input for an output code of 0001, which is 0.625 V, lies at the center of this range.

6.4 QUANTIZATION NOISE

An ideal ADC or DAC has a certain irreducible error which is known as the quantization error and is introduced by the process of quantization of the analog signal. We can see from the transfer functions that no converter can distinguish an analog difference of less than Q. This error can be as large as $\pm Q/2$. The error is shown in Fig. 6.4 and is a triangular function. This error is a source of noise introduced into the signal by the quantization process. The average value of this quantization noise is zero and its r.m.s. value is $Q/\sqrt{12}$. The process of quantization will therefore add noise to the signal that cannot be eliminated but only reduced to a level consistent with the desired accuracy by using a converter of sufficiently high resolution. The signal-to-noise ratio introduced by the converter is, in decibels,

$$\text{S/N (db)} = 20 \log \left\{ \frac{2^n Q}{Q/\sqrt{12}} \right\}$$

$$= 6.02n + 10.8$$

Thus the signal-to-noise ratio increases by 6.02 dB for each additional bit of converter resolution.

6.5 DIGITAL TO ANALOG CONVERTERS (DACs)

DACs are an integral part of data conversion systems. Not only are they used to produce an analog representation of digital data in a system but are also used in some types of analog to digital converter (ADC). We will therefore embark on a brief discussion of ADCs.

Most DACs can be represented by the schematic diagram of Fig. 6.5. The switching network produces an output current or voltage whose magnitude, derived from the reference voltage and resistor array, depends on the digital

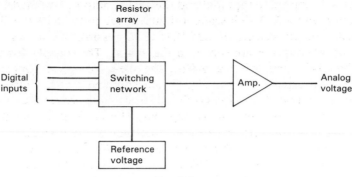

Fig. 6.5

input value. If the output is a current, the amplifier is a transconductance type to produce an analog voltage.

There are several practical implementations of Fig. 6.5 of which one of the most popular (and easy to understand) is the R-2R ladder DAC (Fig. 6.6).

The switches (usually electronic CMOS devices) are controlled by the digital input signal, directing currents either to ground or to the summing line of the amplifier. In Fig. 6.6, we have a four-bit DAC to which a digital word 1011 has been applied, as shown by the switches (0 indicates a line switched to earth). It is relatively straightforward to show that the current at the summing junction is now

$$I_{out} = \frac{V_{ref}}{R}\left(\frac{1}{2} + 0 + \frac{1}{2^3} + \frac{1}{2^4}\right)$$

In general, for an n-bit DAC,

$$I_{out} = V_{ref}\left(\frac{a_1}{2^1} + \frac{a_2}{2^2} + \ldots + \frac{a_n}{2^n}\right)$$

where $a_j = 0$ or 1, depending whether the jth line is at ground or not.

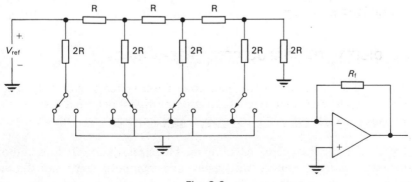

Fig. 6.6

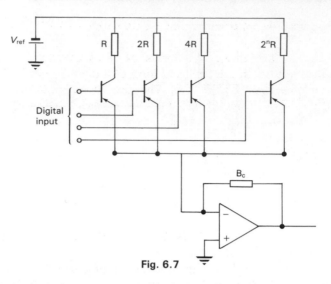

Fig. 6.7

The amplifier at the output converts this current into a voltage proportional to the current. Note that if all digital input bits are set

$$I_{out} = \frac{V_{ref}}{R} \left(\frac{1}{2^1} + \frac{1}{2^2} + \dots + \frac{1}{2^n} \right) = \frac{V_{ref}}{R} \left(1 - \frac{1}{2^n} \right)$$

This type of DAC requires only two values of resistor so that, using laser trimmed thin film resistor networks, matched resistors are easily obtained and the temperature tracking is excellent.

A second popular DAC design is known as the weighted current source. This is shown schematically in Fig. 6.7.

A series of transistors have binary weighted resistors in their emitter leads as shown. The transistors therefore act as binary weighted current sources which can be switched on or off at the digital input terminals. When the digital input at a terminal is high, current flows of value $V_{ref}/2^n$. When the digital input terminal is low, no current flows. The flowing collector currents are added together at the current summing line and can be converted into a voltage by a current-to-voltage amplifier. For greater speed, the output current can be used to drive a load resistor directly. The difficulty with implementing this type of DAC is the wide range of emitter resistors required, with typical LSB resistor values of tens of megohms. Such high values of resistance cause additional problems of temperature stability and switching speed.

6.6 ANALOG TO DIGITAL CONVERTERS (ADCs)

There are far more types of ADCs available than DACs. Of these, about 95 percent belong to one of two types – the integrating type and the successive approximation type.

6.6.1 Successive approximation

Figure 6.8 shows a schematic diagram of the successive approximation ADC. It uses a DAC in the feedback loop of a digital control circuit. The operation of the circuit is as follows. The successive approximation register first turns on the most significant bit of the DAC and the comparator compares this output against the analog input. If the output is greater than the analog input, the bit of the DAC is switched off; if less, the bit is left on. The same procedure is applied to bit 2 of the DAC, and so on until all bits have been tested. At this point, the digital outputs of the successive approximation register represent the converted analog voltage. For an n-bit ADC, n operations are required for a conversion. The conversion efficiency of this type of ADC makes it very fast, and converters with conversion times between 1 and 30 μs are readily available. Figure 6.9 shows the conversion of an analog input for a successive approximation ADC.

There are, however, a greater number of error sources than with the slower, integrating type of ADC discussed in the next section. In particular, the DAC, comparator, and voltage reference are critical components, and can be sources of poor temperature stability.

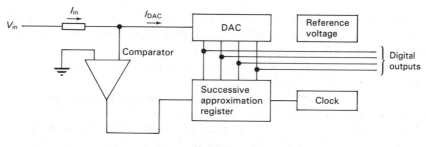

Fig. 6.8

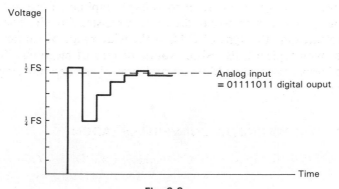

Fig. 6.9

6.6.2 Integrating ADCs

This type does not use a DAC, instead converting the analog input voltage into a time period which is measured. Several variations of the basic theme exist, such as dual slope and charge balancing. They are all relatively slow with conversion times in the range 1–20 ms but have excellent linearity characteristics with noise rejection capability. We shall describe the dual slope system, shown in Fig. 6.10.

The analog input voltage is switched to the input of the integrator and simultaneously the counter begins to count clock pulses for a fixed time T_1 (Fig. 6.11). At the end of this time, the control circuit switches the integrator to the negative reference voltage. The integrator voltage falls and clock pulses are counted for a time T_2 until the comparator detects the integrator voltage falling below zero, when the clock pulses are turned off. Now

$$T_2 = T_1 \frac{V_{in}}{V_{ref}}$$

and so represents the input voltage. The counter value is converted to the required digital word.

By integrating the input, noise can be rejected by this circuit. The accuracy and stability of the system depends only on the reference voltage and integrator circuit linearity.

The accuracy of the dual slope technique depends only on the stability of the clock source (which can be better than one part in a million) and on the value of the reference voltage, which is the only critical component in the

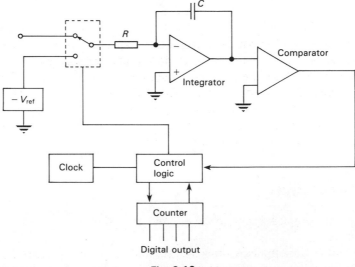

Fig. 6.10

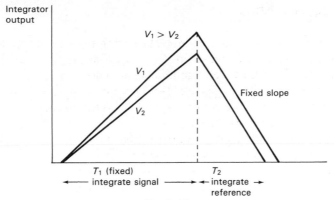

Fig. 6.11

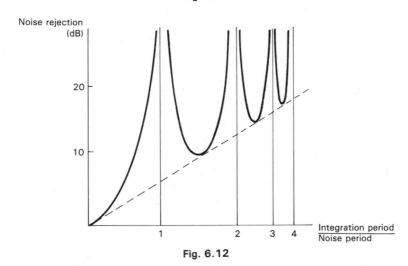

Fig. 6.12

ADC. Changes in other components, such as the integrating capacitor, do not affect the accuracy as long as they do not change during an individual conversion cycle. Since the converter integrates the input signal, the ADC is able to reject noise to some extent. This noise rejection can be very large if the period of the noise is equal to the integrating time T_1, a very useful property if mains interference at 50 Hz is a problem. Figure 6.12 shows the noise rejection capability for integrating type ADCs.

6.6.3 Counter types

These are the simplest types of ADC, and one is shown schematically in Fig. 6.13.

Clock pulses are applied to the counter and the DAC is stepped up one least significant bit at a time. At each step, the comparator compares the

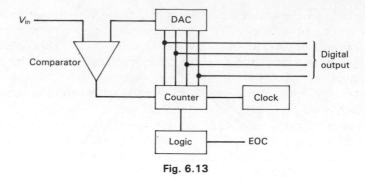

Fig. 6.13

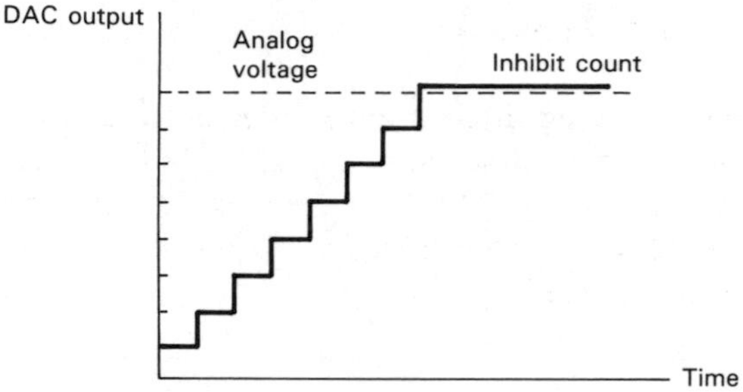

Fig. 6.14

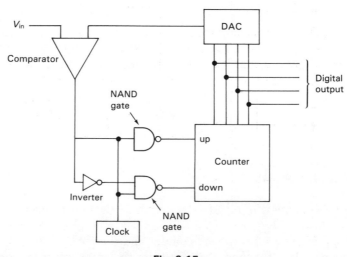

Fig. 6.15

DAC output with the analog input. As soon as the DAC output exceeds the input, counting is inhibited. The counter output is then converted to the required digital word. The steps are shown in Fig. 6.14. This counter is obviously slow and the conversion time depends strongly on the value of the analog input.

Counter type ADCs can be modified to produce the so-called tracking ADC which can continuously follow the input signal, providing continuous digital values of the analog signal. The tracking ADC has an up/down counter controlling the DAC. The output of the comparator is used with some external logic to direct the clock pulses either to the up counter or to the down counter, depending on whether the DAC voltage must increase or decrease to attain the value of the analog input. A schematic diagram of the required logic is shown in Fig. 6.15.

6.6.4 Parallel (flash) conversion

For applications involving digitization of signal containing high frequencies, such as video signal processing, so-called flash ADCs are necessary. Such ADCs work in the 1 to 20 MHz range for eight-bit conversions while speeds up to 100 MHz can be attained with four-bit resolution. The technique is shown in Fig. 6.16.

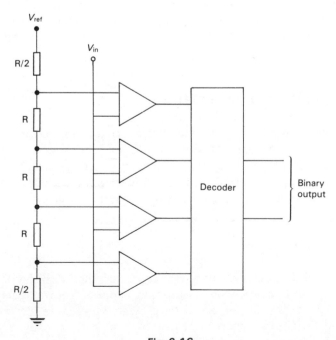

Fig. 6.16

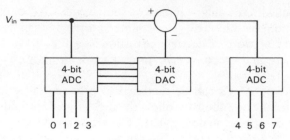

Fig. 6.17

The technique depends heavily on the comparators and their speed of operation. The resistor chain drops the reference voltage by one LSB as we move down the chain. For a given input voltage, all comparators biased below the input switch on while those above switch off. Thus, the input voltage is quantized in one step. The outputs from the comparators is converted to binary form by a high speed decoder.

Parallel ADCs suffer from one major drawback – an n-bit converter requires $2^n - 1$ comparators. For example, a four-bit ADC requires 15 (a reasonable figure) while an eight-bit ADC would need 255! This usually limits practical flash converters to four-bit resolution. However, these four-bit flash ADCs can be cascaded for higher resolution, as shown in Fig. 6.17.

The result of the first four-bit conversion is fed into an ultrafast DAC whose output is subtracted from the analog input. The residue is fed into another four-bit ADC and the two sets of four-bit data represent the eight-bit data.

6.7 ERRORS IN DATA CONVERTERS

We have seen that all data converters introduce a certain error, the quantization error, into a signal. However, there are other sources of errors in real data converters: offset, gain, and linearity errors. In addition, these errors are temperature dependent. In order to specify the accuracy of a data converter, we must take into account these sources of error.

6.7.1 Zero offset error

If zero is applied to the input of a data converter, then zero should appear at the output. In practice, a small output will appear, known as zero offset error. This error is constant over the full scale range. It is illustrated for DACs and ADCs in Fig. 6.18. Most high resolution converters have provision for trimming out this error.

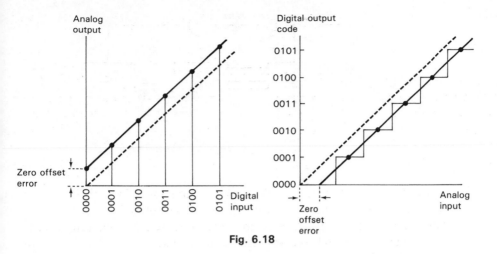

Fig. 6.18

6.7.2 Gain error

In this case, as the input to the converter increases, the output departs more and more from its ideal characteristic. Again, in high resolution converters, provision is made to trim out this error. Figure 6.19 shows this error for DACs and ADCs.

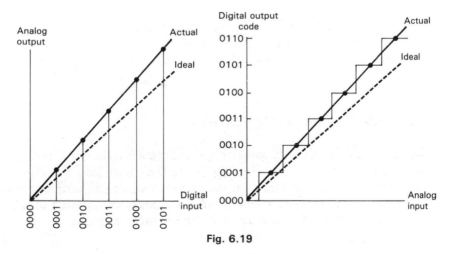

Fig. 6.19

6.7.3 Linearity error

Linearity error is defined with gain and zero offset errors nulled to zero. It cannot be nulled. Two types of linearity error occur in practice.

Integral linearity error is due to the curvature of the input–output function, resulting in a departure from the ideal linear function. It is illustrated in Figure 6.20.

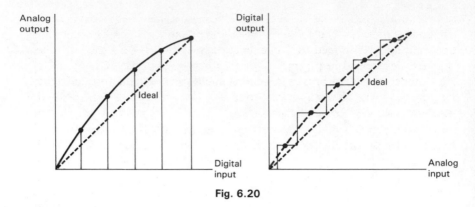

Fig. 6.20

This error is quoted by most manufacturers as the maximum deviation between ideal and actual transfer function.

Differential linearity error is the amount of deviation of any quantum from its ideal value measured between adjacent LSB values. For an ideal converter, this error is zero. Figure 6.21 shows this.

If the differential linearity error is greater than ±1 LSB, then the DAC is said to be nonmonotonic or the ADC is said to possess a missing code. For a successive approximation ADC, differential linearity error is the major source of error.

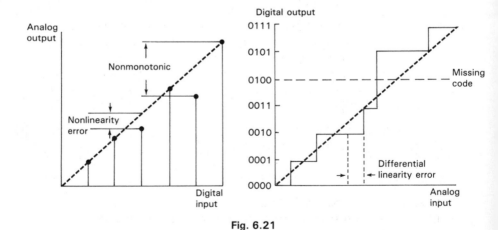

Fig. 6.21

6.7.4 Drift errors in data converters

The errors discussed so far have been effectively considered at a constant temperature. In fact, these errors are temperature dependent. For example, we have seen that both offset and gain errors can be trimmed out but this is only true at one particular temperature. If the temperature changes, the errors will have to be re-zeroed. Linearity errors also change with temperature but these are more troublesome than gain or offset drift since they cannnot be trimmed out. The temperature stability of a data converter is usually expressed in p.p.m./$^\circ$C or fractions of an LSB/$^\circ$C.

If temperature drift proves to be a problem, then several approaches can be adopted. The first is simply to use a converter that has a smaller temperature coefficient. The second approach is to provide a temperature controlled enclosure for the data converter, effectively preventing the temperature of the converter from drifting outside predetermined limits. The third approach relies on the fact that temperature drift in the voltage reference circuit of the converter is a major cause of gain drift. If the converter has provision for an external voltage reference to be used, then the choice of an external reference with a lower temperature coefficient should be made.

Temperature is not the only cause of drift error. Data converters are sensitive to changes in power supply voltages and converter errors due to this source are usually expressed as a percentage of the full scale range for a 1 percent d.c. change in power supply voltage. Finally, converter errors are also affected by changes in internal component values with time. This is usually specified in p.p.m./(time in hours).

6.8 SPECIFIYING CONVERTER ACCURACY

We are now in a position to determine the relative accuracy of the data converter we use in our system. Let us suppose we require eight-bit accuracy, i.e. approximately 0.4 percent accuracy. In a simple-minded fashion we might simply specify an eight-bit converter, taking into account only the quantization error of $\pm\frac{1}{2}$LSB. However, we should take into account the linearity error and drift errors. The following errors may be encountered:

Differential linearity error	$\pm$ 0.2%
Temperature drift error	$\pm$ 0.07%
Power supply drift	$\pm$ 0.002%
Long term drift	$\pm$ 0.02%

The total error will be (including quantization) approximately ± 0.49 percent. This is a worst case error, assuming all errors add in the same direction. Such an error is the quantization error obtained from a six- or seven-bit converter, i.e. the relative accuracy of the converter has been degraded by one or two bits. We would need to specify a 10-bit converter to achieve the desired accuracy.

The figures given above are hypothetical, but illustrate the need for a careful study of the data sheet of the converter for error specifications before deciding the resolution of the ADC needed for the particular application.

6.9 SAMPLE AND HOLD CIRCUITS AND DATA CONVERTERS

An analog to digital converter requires a finite time to carry out a conversion. This conversion time is known as aperture time and can lead to significant errors in a converted signal if the signal is time varying. During the aperture time, t_a, the voltage changes and, if this change is greater than $\pm \frac{1}{2}$LSB, the output digital signal will be in error. We can work out a relation between the aperture time and the rate of change of signal for a given resolution. If the signal to the ADC changes by ΔV during the time t_a, then

$$\Delta V = \frac{dV}{dt} \times t_a$$

If the full scale voltage is FS, then the fractional error is

$$\varepsilon = \frac{\Delta V}{FS} = \frac{t_a}{FS} \frac{dV}{dt}$$

For a sinusoidal signal

$$V = V_0 \sin \omega t$$

which we wish to digitize with an error of $\pm \frac{1}{2}$LSB in 256 (i.e. an eight-bit converter)

$$\varepsilon = \frac{0.5}{2^8} = \frac{t_a}{2V_0} \times \omega V_0$$

since the maximum rate of charge of the signal occurs at $t = 0$ and FS $= 2V_0$. Then

$$t_a = \frac{1}{256} \times \frac{1}{2\pi f}$$

Taking t_a as 10 μs (quite a fast converter) then $f \approx 100$ Hz. Thus only signals of lower frequency than this will be converted within the given error. This figure is quite low and gets worse for higher resolution converters.

The way around this problem is to use a sample and hold converter. We have come across this device in relation to the boxcar. It has two output states. When in sample mode, the voltage at the output is frozen (theoretically indefinitely) at the value of the input when the hold command is given (Fig. 6.22). This value can be converted by the ADC, reducing its effective aperture time considerably.

Of the two commonly used ADCs, sample and hold circuits are only found with successive approximation converters. Integrating converters are only used where a voltage is integrated over a period of milliseconds. Sample and hold circuits typically have aperture times of 50–100 ns and, for an eight-bit converter using a sample and hold with 100 ns aperture time, signals up to 10 kHz can be converted with $\pm \frac{1}{2}$LSB accuracy. A block diagram of a popular sample and hold is shown in Fig. 6.23.

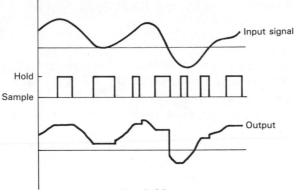

Fig. 6.22

The input amplifier acts as a buffer with a high input impedance. The electronic switch (usually a MOSFET device) is controlled by an input electrode and can be switched rapidly on or off by a driver circuit. When the switch is closed, the voltage on the capacitor C follows the input voltage (sample or track mode); when the switch is opened, the voltage on the capacitor C is held at the value of voltage when the switch was opened (hold mode). In theory, this voltage will remain at this value until the switch is closed again. In practice, this voltage will fall, or droop, with time.

This is specified in the data sheets as the droop rate (usually as $\mu\text{V } \mu\text{s}^{-1}$), which is a temperature-dependent quantity. There are several contributions to the leakage of current in a sample and hold system. There will be leakage through the switch (i_1 in Fig. 6.23), the bias currents from the output amplifier will affect the charge on the capacitor (i_2 in the diagram), while there will be inevitable capacitor insulation leakage (i_3 in the diagram). The total current leakage is $i_1 + i_2 - i_3$ and it should be noted that this can be positive or negative. It follows that the droop rate is

$$D = \frac{dV}{dt} = \frac{1}{C}\frac{dQ}{dt}$$

$$= \frac{1}{C}(i_1 + i_2 \quad - i_3)$$

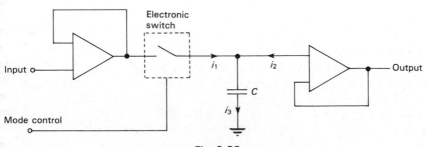

Fig. 6.23

For a system where $i_1 = 25$ pA, $i_2 = 15$ pA and $i_3 = 12$ pA

$$D = 2 \ \mu V \ \mu s^{-1}$$

for a capacitance C of 14 pF.

Another problem in the hold mode is the phenomenon of feedthrough, where, due to capacitive coupling across the switch, a fraction of the input signal will appear at the capacitor. This feedthrough is usually specified in decibels (dB). Droop and feedthrough determine the accuracy of the sample and hold device.

The droop rate can be improved in several ways. The output amplifier should be chosen to have very low bias currents, while the switch should have low leakage. The capacitor C, if an external one has to be supplied, should be of high quality with low dielectric absorption, such as Teflon, polystyrene, or polycarbonate. (Some devices have internally supplied capacitors with terminals for the connection of an external capacitor.)

There are several other parameters specified for the sample and hold device of which the designer should be aware. The first of these is the acquisition time, which is the time required for the sample and hold to reach its final value on receiving a hold command. This time will depend on the slew rate of the amplifiers and also on the 'on' resistance of the switch (RC charging). Because of this latter factor, fast acquisition time systems use small values of capacitor, although this would then imply a large droop rate. To overcome this problem, two sample and hold circuits are needed, the first having a fast acquisition time which then feeds a second sample and hold, which has a large capacitor and hence a low droop rate. The aperture delay is the time required for the switch to open fully after initiation of a hold command. This aperature delay will vary with time and is known as

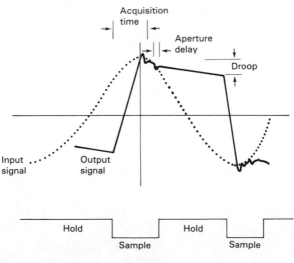

Fig. 6.24

the aperture uncertainty. This uncertainty will introduce an error in the sampled voltage which will increase with the slew rate of the input signal. These factors are illustrated in Fig. 6.24.

6.10 DIGITAL DATA ACQUISITION SYSTEMS

A complete data acquisition system, incorporating several analog channels, is shown in Fig. 6.25. This is a low cost, relatively low speed solution. Each input is sampled in sequence and then converted. It avoids the cost of a separate sample and hold and ADC for each analog channel.

The analog mutiplexer is a device which will accept several input signals, allowing only one of those inputs through to the single output line. A selection code determines which input is gated onto the output. Such a multiplexer, in its simplest state, is a motor-driven commutator switch but modern multiplexers are fabricated from semiconductor switches contained in integrated circuits. Such multiplexers may have control inputs for each switch (i.e. each switch may be actuated by a logic one or zero applied to the control input) or they may have on-chip decoders. Such decoders will accept a binary input signal and actuate one switch dependent on the binary code applied. These devices usually employ complementary MOSFET technology (CMOS) having switching speeds of around 1 μs, although junction FET (JFET) devices can have switching speeds of less than 150 ns.

In using such multiplexers in data acquisition systems there are several factors which can lead to the introduction of errors into the signals. The signal will be attenuated by the semiconductor devices in the transmission path. The FET switch in the 'on' state can be considered as a small resistance (about 50 Ω for CMOS devices and 20 Ω for JFET devices) and a capacitance to earth (about 20 pF for CMOS and 12 pF for JFET). Thus, there will be signal attenuation which will be frequency dependent and can introduce phase shifts at sufficiently high frequencies. For example, in

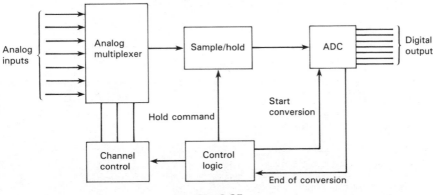

Fig. 6.25

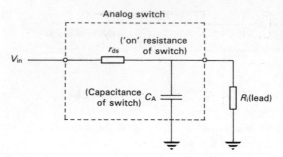

Fig 6.26 Equivalent circuit of FET switch in 'on' state

Fig. 6.26 we have a CMOS device feeding a 10 kΩ load. The d.c. loss is

$$\text{d.c. loss} = \frac{50}{50 + 10^4} \times 100\%$$

$$= 0.5\%$$

The 'off' switches in the multiplexer can affect the signal in the 'on' channel. This effect is known as crosstalk, and is due to leakage currents through the finite (but very high) 'off' resistances of the switches and through capacitive coupling between switches. The leakage currents produce voltages across the load in addition to the signal in the 'on' channel, producing a spurious component at the output (Fig. 6.27).

Since the switches are resistances, they will produce thermal noise which will add to the signal at the output of the multiplexer. For an N-channel multiplexer with one 'on' channel, the system is equivalent to the parallel resistance of the 'on' channel and the N-1 'off' channels. Since the 'on' resistance is very much less than the 'off' resistance, the thermal noise is essentially due to the 'on' channel and will have a value over a 10 kHz bandwidth for a CMOS device of, typically, 0.1 μV at 300 K.

Another source of spurious output occurs when the multiplexer is switched from one channel to the next. Voltage switching transients or spikes can

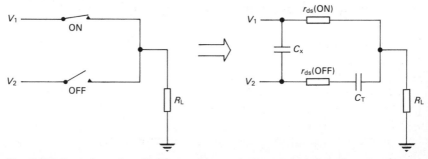

Fig. 6.27 Two channel multiplexer and its equivalent circuit. C_x is the capacitive coupling between ON and OFF switches. C_T is the lumped capacitance of the OFF switch.

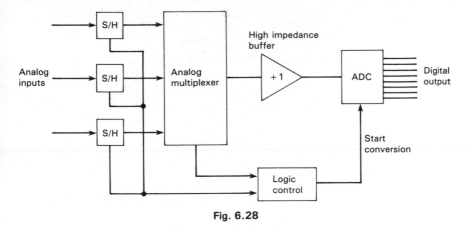

Fig. 6.28

then occur at the output which can add to the signal. For low level signals, the switching transients are the main source of error. In order to minimize their effect, the signal may be preamplified to such a level as to make the switching errors negligible or, alternatively, a filter may be incorporated at the multiplexer output.

A somewhat faster approach (but more expensive) to digital data acquisition systems is shown in Fig. 6.28. There are also greater constraints on this circuit. All the sample and holds must be signaled simultaneously and these systems should have the same aperture time. There will be a spread in the aperture time for each device, known as the aperture uncertainty time, which could limit the maximum rate of change of input signal this circuit can handle. We should also note that the sample and hold devices are sampled sequentially and the last sample and hold in the system must hold its voltage until all other outputs have been converted. This means that the sample and hold must be of high quality with a low droop rate. Notice the use of a high impedance buffer amplifier between the low impedance analog mutiplexer circuit and the ADC.

6.11 VOLTAGE TO FREQUENCY CONVERTERS AS ADCs

Voltage to frequency (v/f) converters are devices which convert an input voltage to a series of digital output pulses whose frequency is directly proportional to voltage. These devices are extremely versatile; since one of their uses is in ADCs, it is an appropriate place to discuss these here. These devices are highly linear (typically 0.002 percent) with excellent temperature stability (typically 10 p.p.m or better).

A simple circuit for the use of a v/f converter as an ADC is shown in Fig. 6.29.

The timer generates a pulse of precisely defined duration and the output of the v/f converter is gated with this pulse. The counter thus registers the

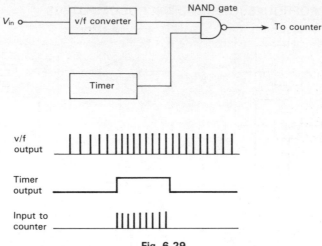

Fig. 6.29

number of pulses during the period of the timer pulse. The resolution of the system is determined by the full scale frequency of the v/f and the pulse width of the timer. For a 1 kHz v/f converter and a 1 second timer pulse, a resolution of ± 1 in 10^3 is possible. By increasing the frequency of the converter, or increasing the width of the timer pulse, we can increase this accuracy. It can be seen that for a given converter, by counting for long enough, a high resolution system can be obtained. Because of this time factor, such ADCs are slow and are usually used in digital voltmeters.

A further drawback occurs when we are measuring small voltages with the ADC since a very long conversion time would be needed for good resolution. This can be overcome by the use of a dual counter scheme and a high frequency clock, as shown in Fig. 6.30. The flip-flop is used to synchronize the clock and v/f pulses. In this case, if counter 1 registers n pulses and counter 2 N pulses, the input voltage is proportional to n/N and the resolution is ± 1 in N.

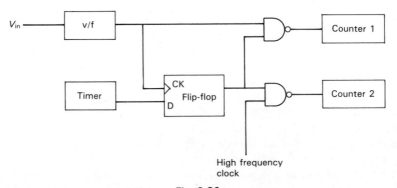

Fig. 6.30

6.12 MICROPROCESSORS AND MICROCOMPUTERS

In recent years the microprocessor has had a dramatic impact in the field of signal processing. In particular, the use of the microprocessor in small, cheap digital computers – the microcomputer – means that most laboratories now have some form of computer control, data logging, and manipulation of data. Systems run from the relatively cheap microcomputers, such as the BBC microcomputer, to more sophisticated systems, such as the IBM PC.

A digital computer is a combination of digital devices that can perform a sequence of operations. This list of operations is known as the computer program. The program is sometimes known as the software of the computer while the combination of the digital devices is known as the hardware. The digital computer contains five essential elements – a control unit, arithmetic logic unit (ALU), a memory unit, an input unit, and an output unit (Fig. 6.31).

The ALU is the place where arithmetic and logical operations are performed on data. The particular type of operation performed is determined by signals from the control unit. The control unit also determines the flow of data between memory and input/output units and the ALU, i.e. the control unit directs the operation of all other units by providing timing and control signals. The control unit and ALU constitute the central processing unit (CPU). Large scale integration has enabled the CPU to be placed on a single silicon chip and it is this chip which is known as the microprocessor. A digital computer containing a microprocessor is known as a microcomputer.

The basic unit of information in digital systems is the binary digit (or bit) which has two states, 0 or 1.

The primary unit of information in the microprocessor is a group of bits, known as a word. The word size is so important that it is often used in describing the microprocessor. Many popular cheap microprocessors use eight-bit microprocessors, of which the most frequently used is the 6502 (in

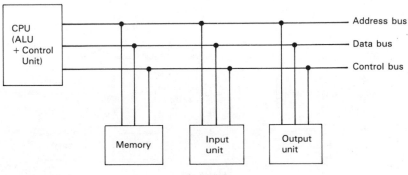

Fig. 6.31

such microcomputers as the PET, Apple, BBC) and the Z-80 (the Spectrum and many S-100 computers, for example). Lately, 16-bit microprocessors, such as the Intel 8086 and Motorola 68000 (which actually performs 32-bit operations internally) have come to the fore. They operate at much higher clock frequencies than eight-bit microprocessors and have much more powerful instruction sets (for example, they are able to multiply and divide; eight-bit microprocessors can only add and subtract so that the user has to generate programs to perform multiplication and division with these chips). However, 16-bit systems are generally more expensive and complex to use than eight-bit systems and the user should look carefully at his particular application before choosing the word size of the microprocessor. Furthermore, each microprocessor type, whether 8- or 16-bit, will have its own internal structure (or architecture), and the detailed operations each is capable of will differ.

One of the major advantages that digital systems have over analog systems is the ability to store large quantities of digital information and data for long and short periods of time. The basic digital storage device is known as the flip-flop, and advances in semiconductor integration have enabled thousands of these flip-flops to be placed on a single silicon chip. These semiconductor memory chips are constructed from either bipolar or MOSFET transistor technology and are the fastest memory devices commonly available. Because of their high speed, semiconductor memory is used as internal (or primary) memory in microcomputers where fast operation is important. Various forms of external (or secondary) memory exist, such as magnetic tape or floppy disks, and these are conventionally used for long term storage of large amounts of data. However, the falling costs of semiconductor memory now make such memory an attractive proposition for secondary memory devices.

Semiconductor memory may be divided into volatile and nonvolatile memory, as shown in Fig. 6.32, which shows further subdivision to be discussed.

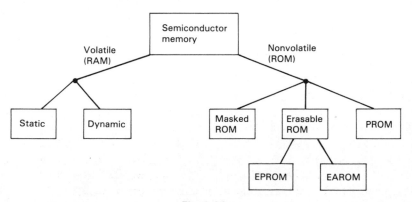

Fig. 6.32

Volatile memory (known as random access or RAM − a slight misnomer) will lose its information when power is removed from the chip. It is possible to read the information available in the RAM or to write new information to it. Static RAM is made up of flip-flops while, in dynamic RAM, information is stored as charge in MOSFET capacitors. Unlike static RAM, the information in dynamic RAM will leak away and so needs to be refreshed approximately every two milliseconds. Dynamic RAM therefore needs more complex external support circuitry than static RAM, but has a higher packing density (more memory cells per chip) and a lower power consumption per memory cell.

Nonvolatile memory (known as read only memory or ROM) retains its information in the absence of power. We can only read information from it and cannot write information to it. Masked ROM has its information permanently written into an integrated circuit as it is being made. The initial cost of such a device is high and is only profitable if large numbers are sold. The programmable ROM (or PROM) can be programmed by the user after purchase. The programming process is irreversible and so must be right first time. Erasable ROMs, however, can be programmed, erased by the user, and subsequently reprogrammed. The most commonly used device, the EPROM, is erased by irradiating the device with ultraviolet radiation. The other device, the electrically alterable ROM or EAROM, can have its memory erased electrically in circuit and promises great future potential.

One particular device seems to lie between these two classifications. These are memories based on CMOS technology, a combination of p-channel and n-channel MOSFETs. They have particularly low power consumption when not being accessed and can be powered from rechargeable batteries as back-up to mains power. When mains power is removed, the battery back-up will enable the information to be retained in the memory for periods of months. More and more systems are using these memories for medium term storage.

In many microcomputer systems, storage space much larger than the available main memory is needed. To get over this, and also to provide long term data storage, magnetic tapes and floppy disk systems are used. In both these devices, an electromagnetic head is used to create areas of magnetization on a magnetic material, corresponding to the digital information, as that magnetic material moves under the head.

Magnetic tape is the least expensive form of long term storage. A magnetic film is deposited on a very thin plastic tape and data bytes are recorded in parallel across the width of the tape perpendicular to the direction of motion of the tape. Floppy disks are disks of mylar coated with a thin magnetic film and contained in a cardboard envelope for strength. The disks are mounted on a rotary drive which spins the disk within the envelope. The main difference between these two devices is the speed at which data can be accessed. With tape storage, we have to pass through the data prior to that that we want in order to access it, which can be quite a lengthy process. With floppy disks, data can be accessed in a random fashion and very rapidly. For this reason, unless costs preclude it, a floppy disk system will be used for secondary storage.

The function of the input and output units is to communicate with and interact between the microprocessor and the world outside the microprocessor system. Input and output devices differ from memory devices because of the variety of systems which exchange data with the microprocessor. Data can be in digital or analog form, may be in the form of voltage or current, and the voltage may be at levels unsuitable for the microprocessor. Devices may be much slower than the microprocessor, or they may be fast but producing data irregularly. We thus need circuitry between the external device and the microprocessor. This interface circuitry is known as an input/output port.

Data can be transferred between a microprocessor and an external device in two possible formats – parallel or serial. In a parallel transfer all the information (or a known fraction of it) is transferred at one time. Since the microprocessor handles data internally in parallel, this is the simplest type of data transfer. In the serial mode, one bit of information is transferred at a time and some means (either hardware or software) must be available to convert a parallel word of data into this serial form. Parallel transfers will proceed far more quickly than serial transfers, but will be more expensive since a greater number of separate wires are needed. Serial transfer will therefore be used if data have to be transmitted over large distances.

The components of the microcomputer system are connected by three groups of wires, which are known as buses. All devices in the microcomputer are connected to these buses. The three buses coming from the microprocessor are the data bus, the address bus, and the control bus. The data bus is a bidirectional bus; that is, data can travel in either direction from, for example, the microprocessor to memory or from the memory to the microprocessor. The number of lines constituting the data bus is the same as the word length of the microprocessor. All information transferred under program control travels on this bus via the microprocessor. The address bus is a unidirectional bus and is used to select or address a location in memory. The number of bits in this bus bears no direct relationship to the word length of the microprocessor. Generally, eight-bit microprocessors such as the Z-80 or the 6502 have 16 address lines and so can directly address 2^{16} (= 65 536) different memory locations. This value is 64 Kbytes of memory (1 Kbyte = 1024) and such eight-bit microprocessors are said to be able to access directly 64 K of memory. Sixteen-bit microprocessors generally have more address lines and so are capable of addressing much larger amounts of memory. For example, the Motorola 68000 has 24 address lines and so can address 16 Mbyte of memory. The control bus is used to synchronize the activity of the separate microcomputer elements, such as whether the microprocessor is reading data from memory or writing data to memory. The details of the control bus vary widely from microprocessor to microprocessor.

The problem with such a bus structure is that the microprocessor can only communicate with one device at a time. If more than one device tries to access the microprocessor bus at one time, bus contention arises between the devices and the result can be catastrophic. To overcome this, modern digital devices are made tristate. This means that, in addition to having the states logic 1 and logic 0, the device has a third state where it has effectively an infinite input impedance. In this state, the device appears to the microprocessor to have disappeared from the bus. The device will have a tristate bus control terminal which can be used to switch the device into or out of the high impedance state. When the microprocessor wishes to access a device, it (in conjunction with some external logic) switches the device out of the high impedance state.

These, then, are the devices which make up the microcomputer. To get it to run, a program is needed. The microcomputer will contain a monitor program which will handle all the basic operations to ensure correct operation, e.g. it will allow you to type in information to the computer and display it on a television screen. To get the computer to do a task, the user must enter a program in a language that the computer can understand. In fact, digital computers can only understand binary digits, and in some of the most primitive microprocessor systems data can be entered in binary format. Usually, however, there is some English-like language in which the user can enter his program, the monitor then converting this into binary format (machine code). We can distinguish between low level and high level languages. Low level languages are known as assemblers and each instruction corresponds to an operation of the microprocessor. High level languages, such as BASIC, Pascal, or Fortran, are generally much easier to use than assembler. One instruction in a high level language may correspond to many operations of the microprocessor. Programs can be written much more quickly in high level languages but the machine code they produce is not always the most efficient for a particular task. In input/output systems this inefficiency can be critical, and now most microcomputer systems using high level languages allow you to write that section of the program concerned with input/output in assembler while writing the rest of the program in a high level language.

For reference, a small microprocessor system, based on a Z-80 microprocessor, is shown in Fig. 6.33. It is not the intention of this book to discuss in detail the design of such a system and the circuit is given here for completeness only. The design and implementation of dedicated control boards such as this are dealt with in numerous books (see, for example, Heffer *et al.* 1981) and will need development tools to implement. The use of commercially available microcomputer systems removes the need for such development, and also provides the user with greater flexibility since external devices such as ADCs can easily be interfaced to them via standard interfaces, which are discussed below.

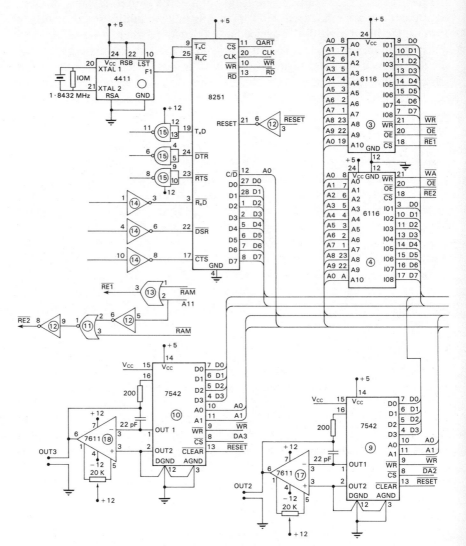

Fig. 6.33 (Courtesy of S. Goodes)

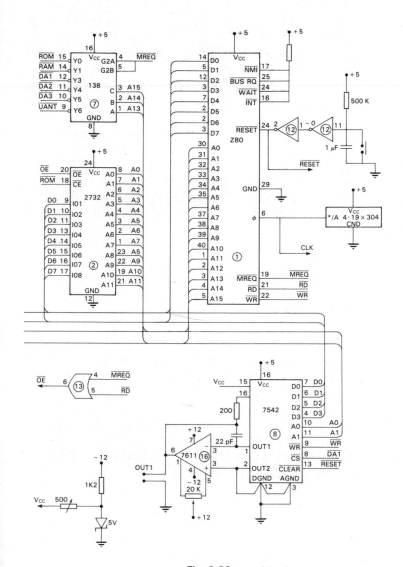

Fig. 6.33 continued

6.13 SYNCHRONIZING DATA TRANSFERS

We now need to manage the transfer of data between the peripheral devices and the microcomputer. There are several reasons for this; for example, most peripherals handle data more slowly than the CPU, or there may be several peripherals which need servicing.

The majority of data transfers between a device and a microcomputer are controlled by a program residing in the microcomputer. We say such transfers are programmed. Such transfers will not support very high data rate transfers, and for this another technique, direct memory access, must be employed.

Programmed data transfers can be subdivided as shown in Fig. 6.34.

Unconditional transfer is the simplest type and assumes the external device is always available and ready for communication with the microprocessor. It is effectively an open-loop process, there being no feedback to the microprocessor from the peripheral to inform the microprocessor that the peripheral is ready to receive data or that data is available for the microprocessor.

Conditional data transfer is often known as handshaking, and communication between the microcomputer and the peripheral is essential. Data is only transferred when both microcomputer and peripheral are ready for such a transfer. The peripheral will signal the microcomputer through status lines, which, as the name suggests, tell the microcomputer the current status of the peripherals. These status lines indicate whether the peripheral is ready for data transfer or busy completing a previous data transfer. In order that the microcomputer may be able to inform the peripheral that it wants to initiate a data transfer, it can produce a data strobe pulse. These lines ensure an orderly flow of data. A schematic diagram of the arrangement is shown in Fig. 6.35.

Suppose we wish to read data from the peripheral. The microcomputer sends out a data strobe to initiate data transfer. The peripheral then responds with a busy level, to indicate that it is fetching data, and this line remains at this logic level until data is available. The ready line now changes

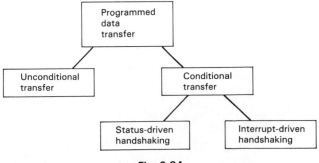

Fig. 6.34

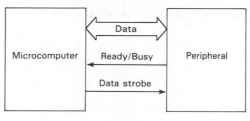

Fig. 6.35

logic state to indicate to the microcomputer that it can now read the data. If we wish to output data to the peripheral from the microcomputer, the microcomputer must check the ready/busy line to see if the peripheral is ready to accept data.

There are two ways of checking if the ready/busy line goes low.

1. *Status driven.* The microcomputer continuously checks the ready/busy line; it is called polling. A flow diagram of the process is shown in Fig. 6.36. It is a technique which is relatively easy to implement, but is relatively inefficient since the microcomputer is tied up with polling.

2. *Interrupt driven.* In this mode, the microcomputer is free to carry on with other tasks until the peripheral is ready to transfer data. The microprocessor has one or more special electrodes called interrupt lines. When a signal appears on this line, the microprocessor stops its current program in an orderly fashion and services the cause of the interrupt. This interrupt signal is derived from the ready/busy line. The data strobe line is used as an interrupt acknowledge signal to indicate to the peripheral that the interrupting peripheral is about to be serviced. At the end of the service routine, the microcomputer returns to the original program at the point it was interrupted (Fig. 6.37).

Interrupts provide the fastest possible response to external devices, but the device is actually serviced by the use of software which may not be quick

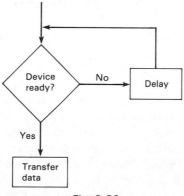

Fig. 6.36

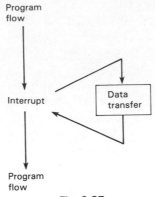

Fig. 6.37

enough for processes involving fast transfers of large amounts of data, for example between main memory and floppy disks. To accomplish such high speed transfers, a specialized chip, known as a direct memory access (DMA) controller, is used. Effectively, this is a specialized microprocessor which is replacing the software transferring data. The controller requires the use of both data and address buses, and, in its simplest form, it suspends the operation of the main microprocessor by sending it into the high impedance state. The DMA controller can then take control of the system and transfer data between two specified devices in that system. The controller usually contains automatic sequencing mechanisms, allowing it to transfer blocks of data without any intervening software.

6.14 INTERFACING MICROCOMPUTERS AND DATA CONVERTERS

In the previous section we have seen how to handle irregular data transfers between external devices such as data converters and a microcomputer. In this section we shall look briefly at the hardware requirements of any interface circuitry.

We have seen that data is transferred between the microcomputer and an external device in a parallel or a serial form. Data converters, of course, transform an external analog signal into a parallel digital form – whether they are of the successive approximation, integrating, or flash type. It follows then that, if high sampling rates are required, data should be transferred in parallel form, which implies that the data converter should be located close to the microcomputer. If, however, the data converter must be placed a long distance away from the microcomputer, the data should be sent in serial form, which will involve a lower sampling rate. Serial transfer of data can involve more complex circuitry than the parallel transfer, which we shall consider first.

The parallel interface will need to possess data latches, which can hold any data stable until it is read either by the microcomputer or the external device, buffers which can drive the data bus, and usually a status register, which the microcomputer can read to indicate if there is data to be read or if it can write data to an external device. This status register is used in the handshaking process. Such a parallel interface can, of course, be built from digital integrated circuits but this can be a complicated task. Fortunately, there exists a device called a programmable peripheral interface (PPI) which is an integrated circuit containing all the components needed for a parallel interface. The device can be programmed (hence its name) for data input, or output, to produce interrupts on certain conditions, produce handshake signals, and so on. Each major microprocessor manufacturer produces its own programmable parallel interface (usually with its own name) and the different devices have slightly different features which should be carefully studied before use, by consulting the manufacturer's data sheets. For example, Intel produce the I8255 device which they call a programmable input/output (PIO) device, which has two eight-bit ports that can be programmed to be input or output and a third port that can be programmed either as a port or as a set of control lines for handshaking. Motorola produce the 6821 peripheral interface adaptor (PIA) which contains two eight-bit ports and four control lines that can be used for handshaking. Two of these control lines can only be used as input lines, while the other two can be programmed to be either input or output (see Figs. 6.38 and 6.39).

A serial interface will contain all the components of the parallel interface but, in addition, it will contain a component to convert parallel data to serial and serial data to parallel. As with the parallel interface, all these components can be found on a single integrated circuit, the UART (universal asynchronous receiver/transmitter). The output of such a device is usually at TTL levels and it is customary to convert this to the so-called RS-232 levels before transmission over long distances. This confers greater noise immunity on the data. In the RS-232 scheme, voltages between -15 V and -3 V are assigned to logic 1 while voltages between $+3$ V and $+15$ V are assigned to logic 0. Similarly, serial data into the UART must be at TTL levels so that incoming data at RS-232 levels must be converted into TTL levels first. Buffer chips to perform such level shifts are readily available. The use of the UART to send data from a remote transducer will therefore involve an ADC to convert the analog data to parallel digital form. The data is then converted into serial form with a UART, producing serial data which is then converted to RS-232 levels before being sent, usually over a twisted pair of cables, to the microcomputer system, where the data is first converted to TTL levels and returned to parallel form by another UART before being read by the microcomputer (Fig. 6.40(a)). An alternative to this, when the data rate is very low, is to use a voltage to frequency converter. In this scheme, illustrated in Figs. 6.40(b) and (c), the analog voltage is converted directly into a stream of pulses which can be transmitted down the twisted cable; at the receiving end the number of pulses

8255A/8255A-5
PROGRAMMABLE PERIPHERAL INTERFACE

8255A OPERATIONAL DESCRIPTION

Mode Selection

There are three basic modes of operation that can be selected by the system software:

Mode 0 – Basic Input/Output
Mode 1 – Strobed Input/Output
Mode 2 – Bidirectional Bus

When the reset input goes 'high' all ports will be set to the input mode (i.e., all 24 lines will be in the high impedance state). After the reset is removed the 8255A can remain in the input mode with no additional initialization required. During the execution of the system program any of the other modes may be selected using a single output instruction. This allows a single 8255A to service a variety of peripheral devices with a simple software maintenance routine.

Operating Modes

MODE 0 (Basic Input/Output). This functional configuration provides simple input and output operations for each of the three ports. No 'handshaking' is required, data is simply written to or read from a specified port.

Mode 0 Basic Functional Definitions:
- Two 8-bit ports and two 4-bit ports.
- Any port can be input or output.
- Outputs are latched.
- Inputs are not latched.
- 16 different Input/Output configurations are possible in this Mode.

MODE 1 (Strobed Input/Output). This functional configuration provides a means of transferring I/O data to or from a specified port in conjunction with strobes or 'handshaking' signals. In mode 1, port A and port B use the lines on port C to generate or accept these 'handshaking' signals.

Mode 1 Basic Functional Definitions:
- Two groups (group A and group B)

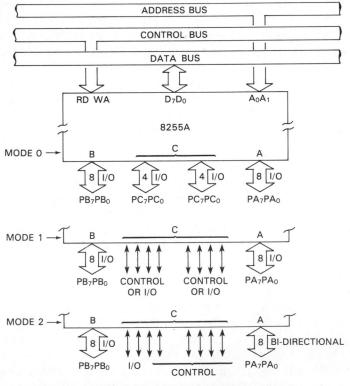

Basic Mode Definitions and Bus Interface

- Each group contains one 8-bit data port and one 4-bit control/data port.
- The 8-bit data port can be either input or output. Both inputs and outputs are latched.
- The 4-bit port is used for control and status of the 8-bit data port.

MODE 2 (Strobed Bidirectional Bus I/O). This functional configuration provides a means for communicating with a peripheral device or structure on a single 8-bit bus for both transmitting and receiving data (bidirectional bus I/O). 'Handshaking' signals are provided to maintain proper bus flow discipline in a similar manner to mode 1. Interrupt generation and enable/disable functions are also available.

The modes for port A and port B can be separately defined, while port C is divided into two portions as required by the port A and port B definitions. All of the output registers, including the status flip-flops, will be reset whenever the mode is changed.

Modes may be combined so that their functional definition can be 'tailored' to almost any I/O structure. For instance: group B can be programmed in mode 0 to monitor simple switch closings or display computational results. Group A could be programmed in mode 1 to monitor a keyboard or tape reader on an interrupt-driven basis.

Single Bit Set/Reset Feature
Any of the eight bits of port C can be Set or Reset using a single OUTput instruction. This feature reduces software requirements in Control-based applications.

Mode 2 Basic Functional Definitions:
- Used in group A *only*.
- One 8-bit, bidirectional bus port (port A) and a 5-bit control port (port C).
- Both inputs and outputs are latched.
- The 5-bit control port (port C), is used for control and status for the 8-bit, bidirectional bus port (port A).

Fig. 6.38   (Reproduced by kind permission of Intel Corporation)

received in a fixed interval of time is counted, or a frequency to voltage converter is used to obtain a measure of the transducer voltage.

Modern data converters now contain many of the elements of both parallel and serial interfaces on-chip and may be connected directly to the microprocessor, but unless you are confident in the design and construction of microprocessor systems, it is probably better to connect them to the microprocessor via a PPI chip or a UART which most microcomputers have as standard. A simple design, employing a buffer amplifier (the high speed, high slew rate OP27), a sample and hold circuit (Analog Devices AD 389), and an ADC (the 12-bit Analog Devices AD 587 successive approximation converter) is shown in Fig. 6.41 connected to a PIA of a microcomputer system.

As noted earlier, the PIA has two eight-bit ports and in this example one port is programmed to be an input (port A) while the other (port B) is programmed so that half the port (PB0–PB3) acts as an input while the other half (PB4–PB7) acts as an output. The signal is fed into the system via an op-amp whose gain is controlled by the analog switch and resistor network as shown. The state of the analog switch is determined by PB4–PB7. By sending one of these lines low under program control, one or more of the resistors can be switched into the feedback path, thus changing the gain under microcomputer control. The control lines of the PIA are designated CA1 and CA2. CA1 is always an input while CA2 is, in this instance, programmed to be an output. CA2 is connected to the start conversion pin of the ADC and an output pulse on this line will instruct the ADC to begin an analog to digital conversion. For this ADC, such a start conversion pulse must be 200 ns long at minimum. Then, 100 ns after the positive

MC6821 (1.0 MHz), MC68A21 (1.5 MHz), MC68B21 (2.0 MHz)

PERIPHERAL INTERFACE ADAPTER (PIA)

The MC6821 Peripheral Interface Adapter provides the universal means of interfacing peripheral equipment to the M680G family of microprocessors. This device is capable of interfacing the MPU to peripherals through two 8-bit bidirectional peripheral data buses and four control lines. No external logic is required for interfacing to most peripheral devices.

The functional configuration of the PIA is programmed by the MPU during system initialization. Each of the peripheral data lines can be programmed to act as an input or output, and each of the four control/interrupt lines may be programmed for one of several control modes. This allows a high degree of flexibility in the overall operation of the interface.

- 8-bit bidirectional data bus for communication with the MPU

- Two bidirectional 8-bit buses for interface to peripherals
- Two programmable control registers
- Two programmable data direction registers
- Four individually-controlled interrupt input lines: two usable as peripheral control outputs
- Handshake control logic for input and output peripheral operation
- High-impedance three-state and direct transistor drive peripheral lines
- Program controlled interrupt and interrupt disable capability
- CMOS drive capability on side A peripheral lines
- Two TTL drive capability on all A and B side buffers
- TTL-compatible
- Static operation

EXPANDED BLOCK DIAGRAM

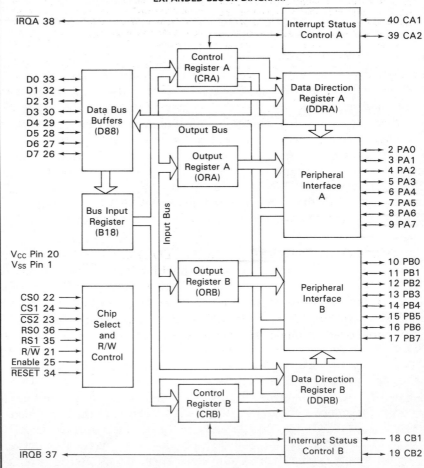

PIA PERIPHERAL INTERFACE LINES

The PIA provides two 8-bit bidirectional data buses and four interrupt/control lines for interfacing to peripheral devices.

Section A Peripheral Data (PA0-PA7) – Each of the peripheral data lines can be programmed to act as an input or output. This is accomplished by setting a '1' in the corresponding Data Direction Register bit for those lines which are to be outputs. A '0' in a bit of the Data Direction Register causes the corresponding peripheral data line to act as an input. During an MPU Read Peripheral Data Operation, the data on peripheral lines programmed to act as inputs appears directly on the corresponding MPU Data Bus lines. In the input mode, the internal pullup resistor on these lines represents a maximum of 1.5 standard TTL loads.

The data in Output Register A will appear on the data lines that are programmed to be outputs. A logical '1' written into the register will cause a 'high' on the corresponding data line while a '0' results in a 'low.' Data in Output Register A may be read by an MPU 'Read Peripheral Data A' operation when the corresponding lines are programmed as outputs. This data will be read properly if the voltage on the peripheral data lines is greater than 2.0 volts for a logic '1' output and less than 0.8 volt for a logic '0' output. Loading the output lines such that the voltage on these lines does not reach full voltage causes the data transferred into the MPU on a Read operation to differ from that contained in the respective bit of Output Register A.

Section B Peripheral Data (PB0-PB7) – The peripheral data lines in the B Section of the PIA can be programmed to act as either inputs or outputs in a similar manner to PA0-PA7. They have three-state capability, allowing them to enter a high-impedance state when the peripheral data line is used as an input. In addition, data on the peripheral data lines PB0-PB7 will be read properly from those lines programmed as outputs even if the voltages are below 2.0 volts for a 'high' or above 0.8 V for a 'low'. As outputs, these lines are compatible with standard TTL and may also be used as a source of up to 1 milliampere at 1.5 volts to directly drive the base of a transistor switch.

Interrupt Input (CA1 and CB1) – Peripheral input lines CA1 and CB1 are input only lines that set the interrupt flags of the control registers. The active transition for these signals is also programmed by the two control registers.

Peripheral Control (CA2) – The peripheral control line CA2 can be programmed to act as an interrupt input or as a peripheral control output. As an output, this line is compatible with standard TTL, as an input the internal pullup resistor on this line represents 1.5 standard TTL loads. The function of this signal is programmed with Control Register A.

Peripheral Control (CB2) – Peripheral Control line CB2 may also be programmed to act as an interrupt input or peripheral control output. As an input, this line has high input impedance and is compatible with standard TTL. As an output is compatible with standard TTL and may also be used as a source of up to 1 milliampere at 1.5 volts to directly drive the base of a transistor switch. This line is programmed by Control Register 8.

Fig. 6.39 (Reproduced by kind permission of Motorola Inc.)

edge of this pulse, the end of conversion (EOC) line goes high and this causes the sample and hold converter to hold the input signal (Fig. 6.41). On the negative-going edge of the output pulse from CA2, the conversion begins. At the end of the conversion EOC goes low, which can be detected on the CA1 control line of the PIA. This also sends the sample and hold circuit into the sample mode. The microprocessor can then read in the 12-bit digital data, store it in memory and then begin the whole process again.

There are two limitations to the rate of digitization signals in this example – throughput rate and aperture jitter of the sample and hold. The ADC completes its converson in 3 μs while the acquisition time of the sample and hold is 300 ns. Thus, a complete conversion and acquisition will take place in about 4 μs, giving a sampling rate in this example of $\simeq 250$ kHz. This will drop to a much lower rate because of the time needed to store data. Aperture jitter on the sample and hold will also limit this rate. If we try to hold

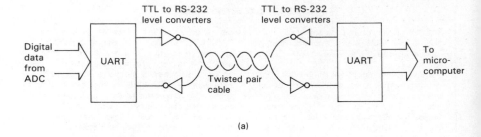

(a)

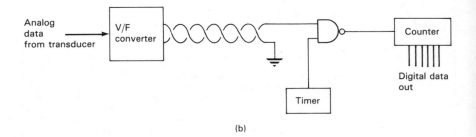

(b)

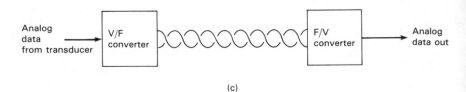

(c)

Fig. 6.40 Serial communications: (a) using UARTs; (b), (c) using voltage-to-frequency converters

a signal at some time, then, due to aperture jitter, the signal may be fully held some time t_a (the aperture jitter) later, during which time the signal may have changed. For a signal

$$E = E_0 \cos \omega t$$

the maximum rate of change of the signal is $E_0\omega$. If the maximum error which can be allowed is $\pm \frac{1}{2}$ LSB then the signal should not change by this amount in the time t_a, i.e.

$$E_0\omega \leqslant \tfrac{1}{2} \, \text{LSB}/t_a$$

Thus for an n-bit converter, the maximum signal frequency is

$$f_{\max} = \frac{2^{-n}}{2\pi t_a}$$

For the AD 389, $t_a = 0.4$ ns and, with $n = 12$; this gives a maximum frequency of 97 kHz.

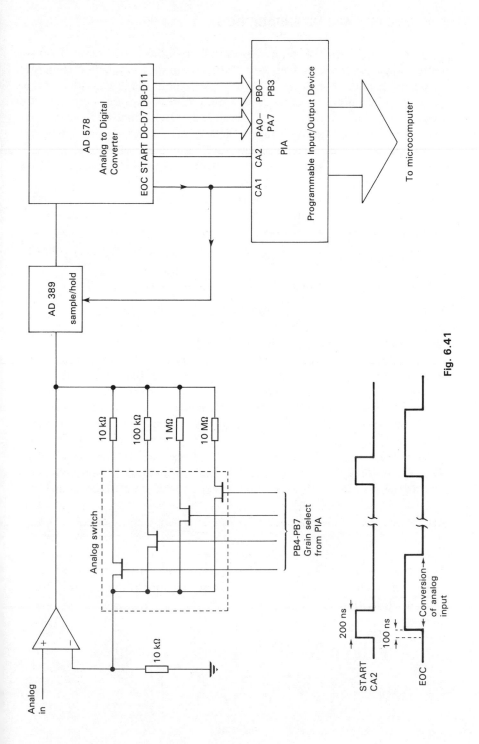

Fig. 6.41

6.15 THE IEEE-488 INTERFACE BUS

The interfacing described so far is essentially a point-to-point system. If we wanted to place several devices onto the microcomputer, it would seem that we would connect each one to its own PPI chip. This can be a rather inelegant solution and a more flexible design involves the use of standard multipoint or bus connections, of which the most common, at least in instrumentation applications, is the IEEE-488 bus. Such a structure involves the simple connection of equipment such as digital voltmeters, printers, plotters, and so on to a microcomputer, without detailed knowledge of how the interface works (Ricci and Nelson 1974).

In this bus structure, the microcomputer is connected to external instruments via a single cable containing 24 wires. The wires have the following functions:

1. Eight ground lines
2. Eight data lines
3. Eight control/management lines

A block diagram of the IEEE-488 bus is shown in Fig. 6.42.

In operation, the bus defines several types of device. The first is a controller, which in this case is the microcomputer. Only one controller is present on the bus. Devices may be talkers, that is, they present data to the bus; they may also be listeners, accepting data from the bus; or finally they may be both talkers and listeners.

The data bus is bidirectional, and to manage and control the flow of data on the bus up to eight other lines are specified (not all these lines will take part in every data transfer). Three of these lines are used to control the communications, i.e. are handshake signals. These are DAV (Data Valid), NRFD (Not Ready For Data), and NDAC (Not Data Accepted). We shall look at the way in which these lines are used in a simple data transfer.

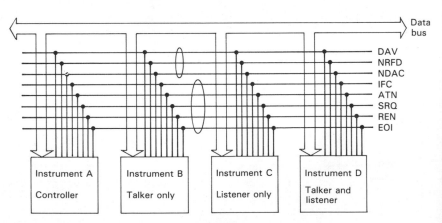

Fig. 6.42

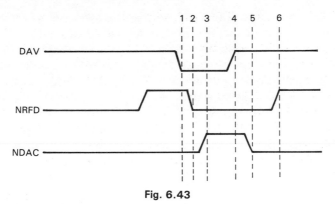

Fig. 6.43

In this data transfer there can be only one talker and one listener on the bus. The talker controls the DAV line, while the listener controls NRFD and NDAC. The listener first raises NRFD to signify that it is ready to accept data. The talker places its data on the data bus and, after allowing time for the data to settle on the bus, lowers DAV to indicate to the listener that valid data is available. When the listener senses this, it lowers NRFD and reads in the data. It then raises NDAC to indicate to the listener that it has accepted the data. The talker can sense this and pull DAV high to tell the listener that there is now no longer valid data on the bus, and the listener responds by pulling NDAC low, acknowledging that data has been removed from the bus. Finally, the listener will pull NRFD high, indicating that it is ready for the next byte of data. (Fig. 6.43).

Five lines are used in the management of the bus. At the beginning of any sequence, the ATN line is pulled low to indicate to devices on the bus that a data transfer is about to occur. The first byte that is placed on the bus is used as an address to identify a talker (only one talker may be on the bus) and the next byte to identify a listener (in fact up, to 14 listeners may be activated). Once this has been achieved, ATN is brought high and then only the talker and the listeners previously addressed may take part in the subsequent data exchange. Thus, each device on the bus must be assigned a unique address which can be set, for example, by means of switches to a value of the designer's choice. To change either talker or listener, ATN must be brought low again. IFC (Interface Clear) is a reset signal which brings all instruments on the bus into a known state. SRQ (Service ReQuest) is a line which tells the controller that a device needs servicing. REN (Remote ENable) must be set low and a device addressed as a listener before it will operate under the command of the bus. EOI (End Or Identify) is a line which is pulled low by a talker during the transfer of its last data byte to signify the end of a message. This line is also used in conjunction with ATN to initiate a parallel poll sequence. In this mode, both EOI and ATN are brought low and an eight-bit data byte read. Each device on the bus pulls one data line low if it is ready, so that the byte read by the controller represents the status of the instruments connected to the bus.

Instruments are connected into the system with a special piggyback cable, which has a single male and female connector at either end that allows one cable to be stacked on top of another. Electrical considerations limit the total number of instruments which can be connected to the bus to 15 and the maximum length of cable to 20 m. A maximum data rate of 1 Mbyte per second is possible on this bus. The drawback of this system is that instruments need some level of intelligence before they can be connected to the bus. However, many manufacturers produce instruments which have IEEE-488 interfaces and most microcomputers will either have an IEEE-488 interface port already built in or one can be bought and easily fitted into the microcomputer. Operating system software is usually provided so that bus control and management occurs automatically without the user having to worry about it, and data can be written to or read from devices having specified addresses quite simply in languages like BASIC.

6.16 THE MICROCOMPUTER AND DATA MANIPULATION

In this and the following sections we shall look at the ways in which the flexibility and power of the microprocessor can be utilized. Raw data may be processed by the microcomputer as it is being produced (real-time processing) or recalled from storage to perform some post-run processing.

6.16.1 Digital averaging

Perhaps the simplest use of the microcomputer is as a signal averager. The basic idea is to take N samples of a given signal, add them together and then divide by the total number of samples, i.e. take the average. As we saw in Chapter 5, this increases the signal-to-noise ratio by a factor of $\sqrt{N}$. A widespread example of this use is in optical and infrared spectroscopy. The spectral range to be scanned by the relevant monochromator is divided up into a suitable number of points. At some particular wavelength, data is obtained, either directly in digital form (from a photon counter where applicable) or digitized by an ADC, and then stored in memory. The microcomputer then moves the monochromator on to the next wavelength, more data is taken and stored in the next location, and so on until the complete spectral region has been covered. If the signal-to-noise ratio is inadequate, the spectral region is scanned again. At any wavelength, the digitized data is added to the memory location corresponding to that wavelength in the previous scan. The average data is then computed and stored in that memory location. Such averaging continues until an acceptable signal-to-noise ratio is achieved. If the nth reading at a particular wavelength λ_i is

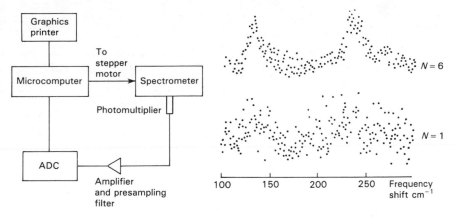

Fig. 6.44

x_n and the value of the memory corresponding to λ_i before storage of this datum is y_n, then the new value stored in this memory location is

$$\frac{\{x_n + (n - 1)y_n\}}{n}$$

An example of this is given in Fig. 6.44. It shows an inelastic light scattering spectrum of a solid with and without signal averaging. The spectrum was taken with a DEC microcomputer and a 10-bit ADC. The data was dumped to a DEC graphics printer. The microcomputer controlled the scanning of the inelastic light scattering spectrometer and at each wavenumber a number of samples determined by the operator was taken, the average being determined by the computer and printed before moving to the next wavenumber. The system was also able to dump the data to a chart recorder via a 10-bit DAC.

A similar example is given by Docchio *et al.* (1980) who discuss the use of such a system to measure the intensity of a pulsed optical source. Details of the acquisition system are given in Fig. 6.45. A trigger pulse for the whole operation is derived from a photodiode facing the pulsed light source and the output from a photomultiplier, used to measure the light intensity, is sent to a gated integrator. The start and end of the integration are both determined by the trigger circuit. The output voltage of the integrator is thus proportional to the energy emitted by the pulsed light source within the monochromator bandwidth. A sample and hold circuit holds the data for digitization by a successive approximation 10-bit converter. Averaging can be performed at a particular wavelength for as many pulses as necessary. The output of the photodiode may also be digitized and the photomultiplier output normalized to the photodiode intensity by software division. Such a process can compensate for short term fluctuations in the lamp intensity. This is an example of the technique used to correct for nonlinearities in experimental systems.

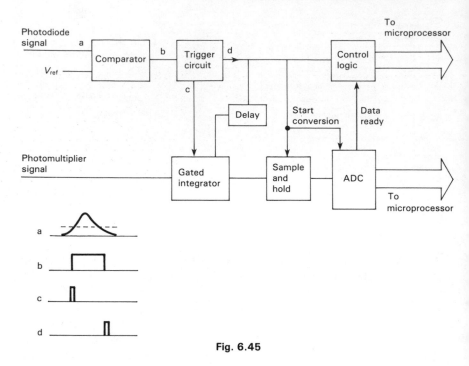

Fig. 6.45

6.16.2 Phase sensitive detection

We have discussed the phase sensitive detector in Chapter 5 and we will not discuss the principles of operation any further. Major drawbacks of analog phase sensitive detectors are the slow measurement times due to long time constants such as that of the output smoothing filters, restricted range of operating frequencies, poor gain and offset stability, low accuracy, and difficulty in reading an analog meter. All these problems can be overcome by using digital techniques.

There have been several schemes published which utilize a microcomputer to produce a digital version of the phase sensitive detector. An example is the design of Momo *et al.* (1981). In this scheme the signal is sampled at two points differing by π for each reference cycle. We have then two samples

$$S_k = x_k = x(kT_0)$$
$$S_{k+\frac{1}{2}} = x_{k+\frac{1}{2}} = x[(k+\tfrac{1}{2})T_0]$$

These two values are subtracted and the operation performed N times, averaging the results. If the signal has the same frequency as the reference, then we have coherent averaging and so have produced a digital version of the analog phase sensitive detector. A schematic diagram of the microcomputer-based phase sensitive detector is shown in Fig. 6.46. An

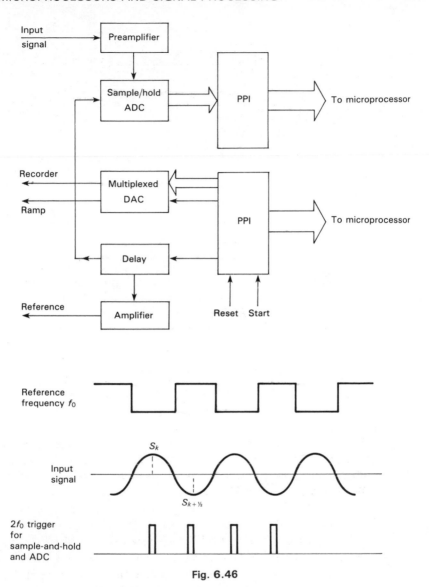

Fig. 6.46

8080-based single-card microcomputer is used, having ROM, RAM, and several programmable parallel interface chips. Data are fed into the microcomputer via a preamplifier, a sample and hold circuit, and an eight-bit ADC. The microcomputer generates a reference frequency f_0 which can be phase shifted from 0 to 2π. This frequency is used to generate a signal at $2f_0$ which is used as a trigger for the sample and hold and the ADC. The timing diagram is also shown in Fig. 6.46. The output of the phase sensitive detector as a function of the frequency difference Δf between the reference and signal frequency is shown in Fig. 6.47. Saniie and Luukkala (1983) have

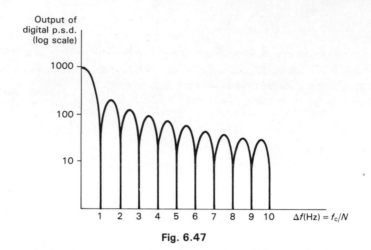

Fig. 6.47

developed a similar microcomputer-based quadrature phase sensitive detector which is capable of producing amplitude and phase information. The system is based on a Z-80 microprocessor with an ADC. The system can be used for digital phase-sensitive detection of signals up to 32 kHz. The sampled data are

$$\text{In phase} \qquad S_k(n) = s(nT + \tau)$$
$$\text{Quadrature} \qquad S_k(n) = s(nT + \tau + T/4)$$

where T is the reference period and τ an arbitrary known delay varying between zero and T.

Iacopini *et al.* (1983) have produced a digital version of the heterodyne phase sensitive detector. However, this version differs somewhat from the above designs in that the synchronous demodulation and integration of the

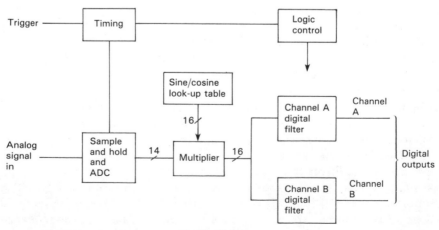

Fig. 6.48

signal take place in digital hardware rather than software, which allows the system to operate at much higher frequencies. A block diagram of the system is shown in Fig. 6.48. At each trigger pulse, the sample and hold and the 14-bit ADC takes one sample of the input and presents the digitized result to the parallel multiplier, where it is multiplied by one point from either a sine or a cosine look-up table stored in a ROM. This produces synchronous demodulation of the signal. The result of this operation is passed into an integrating digital filter (see later) which allows effective time constants of between 20 ms and 5 s.

6.16.3 Digital filters

A particularly important area for digital signal processing is that of digital filtering of signals (Cheetham and Hughes 1982). Digital filters have several advantages over analog filters which help to explain their increasing popularity. Digital filters will show no variation of response with component tolerances, no change in response due to long term ageing of component, and no variations due to temperature effects. Digital filters differ from analog filters due to the sampling process. If the signal is sampled every T seconds, then there will be a faithful reproduction of the signal with frequency components up to $1/T_s$. Thus, the digital filter is considered in the frequency interval

$$\frac{-\pi}{T_s} < \omega < \frac{+\pi}{T_s}$$

The range of digital filter applications is immense, including image and speech processing, telecommunications, and radar, and we shall discuss some of the basic principles of digital filters using only elementary mathematics. An in-depth study of these filters requires a fair amount of mathematics. It is not the purpose of this book to provide such an in-depth study (see, for example, Hamming 1983 for such a treatment) but to acquaint the reader with the scope and applicability of such filters.

Digital filters may be realized in hardware or software. Dedicated digital filter chips are becoming increasingly more readily available and have the advantage of high speed. This means that real-time digital filtering is now available, but such chips are costly. Microprocessors are becoming more powerful and their use in real-time digital filtering is also expanding. Of course, data can also be stored on some mass storage secondary memory device (such as floppy disks) and the digital filtering performed 'off-line'. In this case, speed is not essential and the process can be carried out in software using relatively slow high level languages.

We can gain an understanding of digital filters by studying a simple low pass analog filter, shown in Fig. 6.49(a).

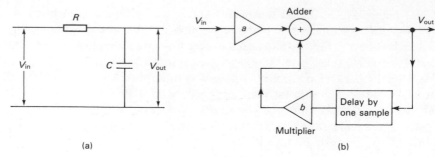

(a)   (b)

Fig. 6.49

The output voltage is given by the solution of the equation

$$RC \frac{dV_{out}}{dt} + V_{out} = V_{in}$$

Let us now convert this to a digital filter by sampling the input signal every T_s seconds, i.e. instead of an analog input voltage V_{in} we produce a series of sampled and quantized voltages $V_{in}(i)$ where $i = 1, 2, \ldots, N$. The filter then produces a series of digital output voltages $V_{out}(i)$. We approximate the differential as

$$\frac{dV_{out}}{dt} \approx \frac{V_{out}(i) - V_{out}(i-1)}{T_i - T_{i-1}} = \frac{V_{out}(i) - V_{out}(i-1)}{T_s}$$

Our original differential equation is now transformed into a difference equation

$$\frac{RC}{T_s} [V_{out}(i) - V_{out}(i-1)] + V_{out}(i) = V_{in}(i)$$

With a little rearrangement, this becomes

$$V_{out}(i) = V_{out}(i-1) + \left(\frac{T_s}{T_s + RC}\right) \cdot [V_{in}(i) - V_{out}(i-1)]$$

$$= aV_{in}(i) + bV_{out}(i-1)$$

Such a difference equation can be solved by using the current value of the voltage together with the previous value of the filtered output to produce the current, digitally filtered output. Figure 6.49(b) shows the digital filter in schematic form. The elements needed are an adder, a mutiplier, and a delay (or memory). This particular digital filter is known as a recursive filter since it uses previous values of the filtered output to calculate the present value. Such filters are also referred to in the literature as infinite impulse response filters.

Although the general procedures for obtaining the frequency response of the filter are complicated, we can obtain the response for this filter since we know that the steady state response to a sinusoidal input is a sinusoidal output of identical frequency but different amplitude and phase. Thus, we take input samples

$$V_{in}(n) = A_i \exp(j\omega n T_s) \qquad j = \sqrt{-1}$$

and, since the difference equation is lower, the output can be written

$$V_{out}(n) = A_0 \exp(j\omega n T_s)$$

where, in general, A_0 will be a complex quantity.

Thus, substituting in the difference equation

$$A_0 \exp(j\omega n T_s) = aA_i \exp(j\omega n T_s) + bA_0 \exp[j\omega(n-1)T_s]$$

and

$$\frac{A_0}{A_i} = \frac{a}{1 - be^{-j\omega T_s}}$$

The amplitude and phase of the frequency response of this filter is

$$\left| \frac{V_{out}(i)}{V_{in}(i)} \right| = \frac{a}{(1 + b^2 - 2b \cos \omega T_s)^{\frac{1}{2}}}$$

$$\tan \phi = \frac{b \sin \omega T_s}{b \cos \omega T_s - 1}$$

We can plot these values as the analog frequency varies from 0 to the Nyquist value, $\omega = 2\pi/2T_s = \pi/T_s$ (Fig. 6.50), for given values of a and b.

A comprehensive treatment of such digital filters should involve the z-transform; it is not intended to discuss this here and the interested reader should consult advanced texts on this topic.

In general, a digital filter can be represented by the equation

$$V_{out}(i) = \sum_{n=0}^{M} a_n V_{in}(i-n) + \sum_{j=0}^{N} b_j V_{out}(i-j)$$

A recursive filter is one where not all the b's are zero. If all the b's are zero,

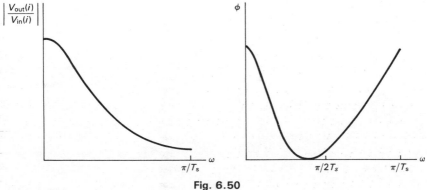

Fig. 6.50

then we have a nonrecursive filter (sometimes known as a finite impulse response filter, or FIR filter). A particularly widely used nonrecursive filter is

$$V_{out}(n) = \frac{1}{\alpha} \sum_{k=0}^{+N} V_{in}(n - k)$$

α is usually set equal to the number of samples of V_{in} used (i.e. $\alpha = N + 1$) and the filter is known as a moving average filter. Let us suppose that

$$V_{out}(n) = \tfrac{1}{10}[\, V_{in}(n - 10) + V_{in}(n - 9) + \ldots + V_{in}(n)]$$

Taking an input of the form

$$V_{in}(n) = A_i \exp(j\omega n T_s) \qquad j = \sqrt{-1}$$

and an output

$$V_{out}(n) = A_0 \exp(j\omega n T_s)$$

then, substituting into the difference equation, we get

$$\frac{A_0}{A_i} = \tfrac{1}{10} \exp(-j\omega T_s) \cdot \frac{\sin\left(\dfrac{10\omega T_s}{2}\right)}{\sin\left(\dfrac{\omega T_s}{2}\right)}$$

The amplitude of the transfer function has the form shown in Fig. 6.51. This is obviously a low pass filter. The effect of the moving average filter on a data set is shown in Fig. 6.52. Note the start-up transient due to the lack of data values until a certain number of samples has been taken.

In general, nonrecursive filters will have a better phase response than recursive filters and are inherently stable since they do not involve feedback from the output of the filter to its input. Nonrecursive filters are simpler to understand and to design. However, nonrecursive filters generally require

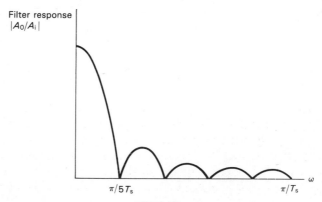

Fig. 6.51

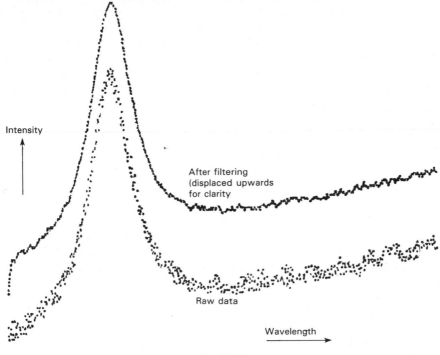

Intensity

After filtering
(displaced upwards
for clarity

Raw data

Wavelength

Fig. 6.52

more computation and/or more digital components than recursive filters and are more likely to be used in real-time filtering situations. In general, however, recursive filters are only used in situations where it is not possible to perform the task with a nonrecursive filter.

The design techniques used for the two digital filter types are quite different (Lynn 1982). Because of the similarities of the transfer function of the recursive digital filter to that of the analog filter, the recursive filter design process is usually a digital version of the analog design. The technique is then to design an analog filter corresponding to the system requirements and to construct a mapping procedure to transform the analog transfer function to the digital one. This procedure is most useful in the design of standard filters, such as low pass, high pass, and band pass. Two techniques are used for nonrecursive filters. They both rely on the fact that the difference equation for these filters is closely related to a convolution operation. The first method is a direct filtering method and involves the direct computation of the convolution of the input signal with some time function (representing the filtering action) to produce the filtered output. The second method operates in the frequency rather than the time domain and involves the Fourier transformation of the input signal. Such a technique is more economical on computing time than the direct convolution method. Once the filter has been designed, especially if it is to be implemented in hardware, the performance of that filter should be modeled in software and its transfer function displayed. This is particularly important for hardware design since, as we shall see below, the use of finite word length arithmetic in hardware design can degrade the filter response. The software simulation can model the effects of the finite wordlength, the effects of the errors can be shown up, and, if necessary, the filter coefficients modified to produce an acceptable response (see Fig. 6.53 for an example of the effects of errors).

As we have stated earlier, digital filters can be implemented in real time using microprocessors. However, the microprocessor does suffer from some shortcomings in these applications, in particular its relatively low speed for performing arithmetic operations. Most eight-bit microprocessors do not have multiplication operations and these multiplications have to be performed using software routines which can take hundreds of microseconds. Since the microprocessor has to take signal samples as well as performing multiplications and additions before finally presenting the filtered value to some device, the whole procedure can take milliseconds, thus restricting the filter to sampling rates of a few hundred hertz. It is possible to increase this rate by special optimization techniques (Tan and Hawkins 1981), but the use of 16- or 32-bit microprocessors with their higher speeds and enhanced instruction sets enable sampling speeds of several kilohertz to be attained (Nagle and Nelson 1981). For example, using a Motorola 68000 microprocessor, sampling rates of up to 5 kHz have been realized for an eighth-order filter. In addition to speed problems, there is a problem of the loss of accuracy through the use of fixed point arithmetic when using microprocessors (with finite word lengths) in digital filters.

Digital signal processing requires that analog values be represented in a binary format with a finite length. This applies to all areas of digital signal processing, including digital filtering, and the finite word length will have several different effects. The basic error of quantization when the analog signal is digitized has already been discussed in detail and will not be discussed further. Some other problems will be discussed in some detail (Oppenheim *et al.* 1972).

Even when we have finite length data representation, the act of processing will produce data which requires extra bits for their representation. For example, a frequent process in digital signal processing is multiplication. If we multiply an n-bit number by another n-bit number, the product will be a $2n$-bit number. This is a problem, since, for example in a recursive digital filter design, frequent iteration will produce a result whose number of bits can increase indefinitely. This can be overcome by treating numbers as fixed point fractions. In this way, the product of two fractions will always remain a fraction and the limited register length can be maintained by truncation.

Fixed point fractions can be easily understood. Consider a four-bit fixed

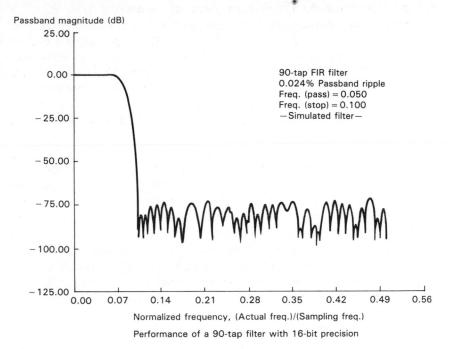

Passband magnitude (dB)

90-tap FIR filter
0.024% Passband ripple
Freq. (pass) = 0.050
Freq. (stop) = 0.100
—Simulated filter—

Normalized frequency, (Actual freq.)/(Sampling freq.)

Performance of a 90-tap filter with 16-bit precision

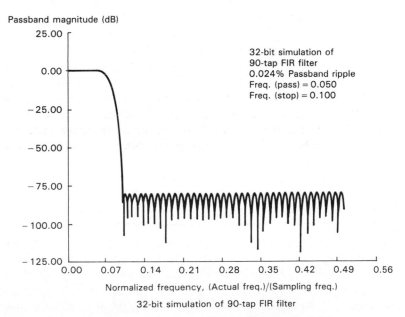

Passband magnitude (dB)

32-bit simulation of
90-tap FIR filter
0.024% Passband ripple
Freq. (pass) = 0.050
Freq. (stop) = 0.100

Normalized frequency, (Actual freq.)/(Sampling freq.)

32-bit simulation of 90-tap FIR filter

Fig. 6.53 (Reproduced by kind permission of Analog Devices)

point fraction $a_0 a_1 a_2 a_3$. This will have a decimal value

$$a_0 \times 2^{-1} + a_1 \times 2^{-2} + a_2 \times 2^{-3} + a_3 \times 2^{-4}$$

Since we can have positive or negative fractions, we use the most significant bit as a sign bit, a 1 in this position representing a positive fraction, a 0 representing a negative fraction. In this representation, data values lie in the range $+1$ to -1.

Let us consider the product of two four-bit signed fixed point fractions

$$1110 \quad \text{and} \quad 0101$$

In decimal terms, these numbers are $+3/4$ and $-5/8$. The eight-bit product of these two fractions is 00111100. Since we are dealing with signed four-bit quantities, we must round this to 0100. We are still left with a fractional quantity, but should notice that the rounding procedure does produce an error. This is known as round-off noise.

Unfortunately, digital signal processing involves addition as well as multiplication. While multiplication of fractions produces fractions, the results of successive additions can produce a sum which is eventually not a fraction. This effect is known as overflow and is usually handled for fixed point arithmetic by requiring that the input data be sufficiently small for overflow to be avoided. The effect of this is to restrict the dynamic range of signals that can be handled.

A third effect of fixed world length is inaccuracies in parameter values. We have seen that the coefficients in digital filters are initially specified with unlimited accuracy, although they can only be implemented with finite word length. Two general approaches have been adopted in dealing with inaccuracies in parameter values. The first is to develop design procedures which are inherently insensitive to parameter inaccuracies. The second is to choose parameter specifications which are consistent with the limited word length. Figure 6.53 shows how the finite word length can degrade the response function of a digital filter and limit the signal-to-noise ratio attainable.

The effects of such arithmetic errors can be avoided to some extent by the use of floating point arithmetic. In this scheme, a number is represented as

$$F = 2^c \times M$$

where M is known as the mantissa (a fraction between 0.5 and 1) and c, known as the characteristic, can have positive or negative values. The product of two floating point numbers is performed by multiplying the mantissa as fixed point fractions and adding the characteristics. The sum of two floating point numbers is obtained by scaling the smaller of the numbers until the characteristics of the two numbers are equal and then adding the mantissa. Floating point arithmetic can introduce errors due to rounding in both multiplication and addition, but its chief benefit is that it provides a much greater dynamic range than fixed point arithmetic. Many manufacturers are now producing chips (often called coprocessors) which can easily be built in to microprocessor systems allowing them to operate with floating point arithmetic. Computers programmed in a high level language usually work with floating point arithmetic, so that if the digital signal processing can be done 'off-line' a high level language should be employed if the advantages of floating point arithmetic are desired.

6.16.4 The fast Fourier transform (FFT)

An alternative method of digital filtering involves the transformation of the sampled signal from the time domain to the frequency domain and then to remove the unwanted frequency components by the application of a suitable filter function. Such a technique involves the use of Fourier transforms and their implementation on digital computers by means of the fast Fourier transform (FFT) (Bergland 1969).

We have come across the Fourier transform before. If we have some continuous function $F(t)$, then $F(t)$ can be written

$$F(t) = \frac{1}{2\pi} \int_{-\infty}^{+\infty} F(\omega)e^{j\omega t}\, d\omega$$

where

$$F(\omega) = \int_{-\infty}^{+\infty} F(t)e^{-j\omega t}\, dt$$

where ω is an angular frequency.

Since we are analyzing the waveform $F(t)$ on a digital computer, it is the discrete version of the Fourier transform (DFT) we must employ. The DFT pair that applies to our sampled waveform, corresponding to the continuous Fourier transform pair given above, is

$$F(k) = \sum_{j=1}^{N-1} F(j)^{2n\,ijk/N} \qquad i = \sqrt{-1}$$

$$F(j) = \frac{1}{N} \sum_{k=0}^{N-1} F(k)^{2\pi\,ijk/N}$$

The computation of the DFT takes a large amount of time. The fast Fourier transform (FFT) is a method for efficiently computing the DFT of a time series. The FFT algorithm is based on the fact that the calculation of the coefficients of the DFT can be carried out iteratively. It was first developed by Cooley and Tukey (1965). If there are $N = 2^m$ samples, the DFT would involve about N^2 operations while the FFT would involve $N \log_2 N$ operations. For example, suppose there are 1024 samples (i.e. $N = 2^{10} = 1024$). Then, the FFT would involve $10 \times 1024 = 10\,240$ operations while the DFT would take 1 048 576 operations. This represents a reduction in computation of greater than a factor of 100. The FFT is readily available on digital computers, even on the smallest systems. A simple FFT for the BBC microcomputer is given by Larsen and Dyrik (1985) and such software is readily available commercially.

The Cooley–Tukey algorithm can be illustrated briefly. For a fuller description, refer to the paper by Cochran *et al.* (1967). The first step is to divide the original series of N samples into two series, $Y(k)$ and $Z(k)$. Each series now contains $N/2$ samples, $Y(k)$ containing the even numbered points and $Z(k)$ the odd numbered points.

$$Y(k) = F(2k)$$

$$k = 0, 1, 2, \ldots, \frac{N}{2}$$

$$Z(k) = F(2k + 1)$$

Now $Y(k)$ and $Z(k)$ are a series of $N/2$ points each, so will have DFTs

$$B(r) = \sum_{k=0}^{N/2-1} Y(k)\exp\left(\frac{4\pi irk}{N}\right)$$

$$r = 0, 1, 2, \ldots, \frac{N}{2} - 1$$

$$C(r) = \sum_{k=0}^{N/2-1} Z(k)\exp\left(\frac{4\pi irk}{N}\right)$$

The DFT $F(k)$ can be written as the sum of the odd and even numbered points

$$F(r) = \sum_{k=0}^{N/2-1} \left\{ Y(k)\exp\left(\frac{4\pi irk}{N}\right) + Z(k)\exp\left(\frac{2\pi ir[2k+1]}{N}\right) \right\}$$

where $r = 0, 1, 2, \ldots, N - 1$.

Or, in terms of $C(r)$, $B(r)$

$$F(r) = B(r) + C(r)\exp\left(\frac{2\pi ir}{N}\right), \qquad 0 \leqslant r < N/2$$

For r greater than $N/2$, there is a periodic repetition of the values for $r < N/2$, so that we may substitute $r + N/2$ for r, i.e.

$$F\left(r + \frac{N}{2}\right) = B(r) + C(r)\exp\left(\frac{2\pi i}{N}\left(r + \frac{N}{2}\right)\right)$$

$$= B(r) - C(r)\exp\left(\frac{2\pi ir}{N}\right), \qquad 0 \leqslant r < N/2$$

The two equations for $F(r)$ and $F(r + N/2)$ allow us to calculate the first $N/2$ points of the DFT of $F(k)$ simply from the DFT of $B(k)$ and $C(k)$. Since these latter functions contain $N/2$ points, each DFT will take $(N/2)^2$ operations, so already we have reduced the total number of computations in the DFT of $F(k)$ from N^2 to $2 \times (N/2)^2 = N^2/2$ operations. We can continue with this operation by dividing $Y(k)$ and $Z(k)$ each into two functions each of $N/4$ points. If $N = 2^m$ this process will continue until we have N one-point transformations. The DFT of a one-point function is the sample itself. It takes $\log_2 N$ of these splittings, so that generating the N-point transform takes a total of about $N \log_2 N$ operations. The discussion has so far assumed that N must be a power of 2, but algorithms have been developed for other N (Singleton 1969). However, if N is restricted to a power of 2, the length of the FFT program is considerably shortened.

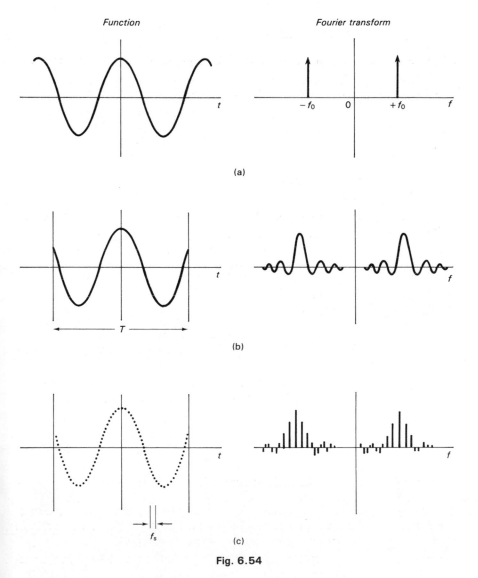

Function   *Fourier transform*

(a)

(b)

(c)

Fig. 6.54

In the interpretation of the DFT, care must be taken since the DFT is performed by looking at the actual signal for T seconds and ignoring everything that happens before or after this 'window'. Effectively, the signal has been multiplied by a rectangular time function (the window) which is equivalent to performing a convolution in the frequency domain. The effect of this is illustrated in Fig. 6.54 (after Higgins 1976). In (a), we see a simple sine wave and its Fourier transform, delta functions at $\pm f_0$. The rectangular window has a transform

$$G(\omega) = \frac{T \sin(\omega T/2)}{(\omega T/2)}$$

for a window of duration T. The convolution of this with the sine wave is shown in (b). We can see that the window has broadened the delta functions and produced a series of spurious peaks or side lobes. Finally, (c) shows the DFT producing aliases which, if the data is sampled at a high enough rate, will be well separated.

The sidelobes can be reduced by applying different windows from the rectangular ones, such as triangular or parabolic windows, although reduction of the sidelobes in these cases is produced at the expense of broadening of the peaks (see Higgins 1976, for some of the most popular window functions).

Having computed the DFT of a signal, the signal can be filtered by multiplying the transformed coefficients by some filter function specified in the frequency domain, i.e. produce the product

$$I(k) = H(k)X(k)$$

where $H(k)$ represents the digital filter function.

There is a considerable degree of freedom in choosing the form of the filter function. However, as above, some caution must be exercised. Consider, for example, the implementation of a digital low pass filter. In frequency space, the simplest function to choose would be a rectangular one (Fig. 6.55).

However, the multiplication in the frequency domain represents a convolution in the time domain. The filter function has a $(\sin x)/x$ form in the time domain and the sidelobes of this function will fold back into the time domain.

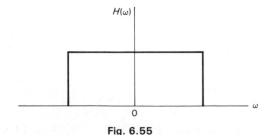

Fig. 6.55

These sidelobes result in aliasing in the time domain. This is because we have looked at the signal for T seconds and neglected everything that has happened before or after. The filter function has been undersampled in the frequency domain. This is overcome by using a filter whose time response dies out or can be truncated (Bergland 1969).

6.16.5 Correlation techniques

Correlation is a technique which can be applied successfully using digital techniques. Autocorrelation is the relation between a signal and the time-shifted version of itself. Let us suppose we have some signal $n(t)$ which we sample every Δ seconds for some period T. The autocorrelation function of the signal is

$$R(t) = \frac{1}{M} \sum_{j=1}^{M} n(j\Delta)n(j\Delta + \tau)$$

Here $\tau = m\Delta$, where m is an integer. The autocorrelation function is computed for all values of the delay time τ, i.e.

$$R(0) = \frac{1}{M} \sum_{j=1}^{M} n(j\Delta)n(j\Delta)$$

$$R(\Delta) = \frac{1}{M} \sum_{j=1}^{M} n(j\Delta)n(j\Delta + \Delta)$$

$$R(2\Delta) = \frac{1}{M} \sum_{j=1}^{M} n(j\Delta)n(j\Delta + 2\Delta) \text{ etc.}$$

A graph of $R(\tau)$ is then plotted against τ.

The autocorrelation function can be used to sift signal from noise. This occurs because the autocorrelation function of a periodic function is itself a periodic function (although with loss of all phase information) while that of random noise of bandwidth B and mean square value N looks like Fig. 6.56.

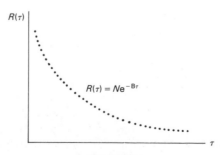

Fig. 6.56

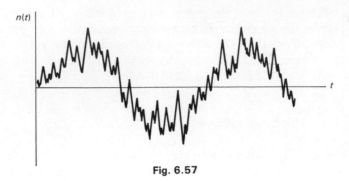

Fig. 6.57

We can see from this that the larger the bandwidth of the noise, the more rapidly the autocorrelation function decays. This is because higher bandwidth noise has higher frequency components so that the noise varies rapidly.

Let us consider a sine wave buried in noise (Fig. 6.57). The autocorrelation of this waveform is

$$R(\tau) = Ne^{-B\tau} + \frac{A^2}{2} \cos \omega\tau$$

which looks like Fig. 6.58.

It can be seen that, at large enough τ, we have recovered the original signal but without the phase information. Thus, we have effected signal recovery without any reference signal or time marker, e.g. in measuring a weak pulse.

Cross-correlation is a similar technique; it measures the correlation between two periodic functions

$$R_c(\tau) = \frac{1}{M} \sum_{j=1}^{M} n_1(j\Delta)n_2(j\Delta + \tau)$$

where $n_1(t)$ and $n_2(t)$ are the two periodic functions. If these two signals have a common fundamental frequency then the cross-correlation function is another periodic function of the same fundamental frequency. $R_c(\tau)$ can

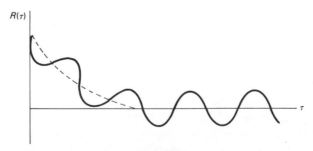

Fig. 6.58

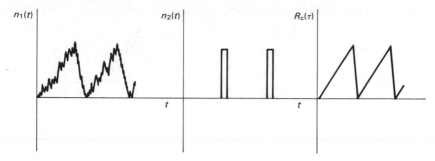

Fig. 6.59

be regarded as a measure of the similarity between two signals. If one of the signals is wideband noise, then $R_c(\tau)$ is zero. This signal recovery technique is even more efficient than autocorrelation but it needs a trigger pulse of the correct frequency. The phase sensitive detector is a special case of a cross-correlator; in the cross-correlator, the 'reference' square wave can be as noisy as you like, but in the PSD it must be 'clean'. An example of such signal recovery is shown in Fig. 6.59.

As a practical example of the use of cross-correlation techniques to enhance the signal-to-noise ratio, Fig. 6.60 shows the enhancement of an infrared spectrum taken from Lam *et al.* (1982). The infrared spectrum was cross-correlated with a 'clean' Gaussian signal and the resulting spectrum showed considerable enhancement of the signal-to-noise ratio.

Correlation functions can also be obtained by using the FFT. Consider two sampled waveforms $n_1(k\Delta T)$ and $n_2(k\Delta T)$. Their DFT pair are

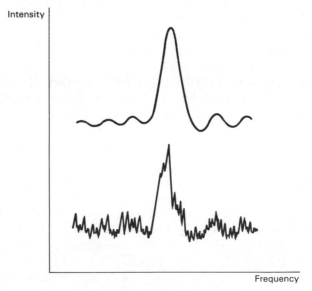

Fig. 6.60

$n_1(k)$, $N_1(j)$ and $n_2(k)$, $N_2(j)$. Consider now the product

$$R(j) = N_1(j)N_2(j)$$

Then we can write

$$r(k) = \sum_{j=0}^{N-1} [N_1(j)N_2(j)] e^{2\pi i jk/N}$$

$$= \sum_{j=0}^{N-1} \left\{ \left[\frac{1}{N} \sum_{\tau=0}^{N-1} n_1(\tau) e^{-2ij\tau/N} \right] \left[\frac{1}{N} \sum_{\tau=0}^{N-1} n(\tau) e^{-2ij\pi/N\tau'} \right] \right\} \times e^{2\pi i jk/N}$$

$$= \frac{1}{N} \sum_{\tau=0}^{N-1} \sum_{\tau'=0}^{N-1} n_1(\tau) n_2(\tau') \left\{ \frac{1}{N} \sum_{j=0}^{N-1} \exp\left[\frac{2\pi i}{N} j(k - \tau - \tau') \right] \right\}$$

Because of orthogonality relationships, this last equation is zero unless

$$k - \tau - \tau' = 0$$

Thus

$$r(k) = \frac{1}{N} \sum_{\tau=0}^{N-1} n_1(\tau) n_2(k - \tau)$$

which is the autocorrelation function if $n_1 = n_2$ and is the cross-correlation function otherwise; the preceding argument has assumed that n_1 and n_2 are periodic functions. This problem is avoided by defining, for example

$$g'(k) = g(k), \qquad 0 \leqslant k \leqslant N$$
$$= 0, \qquad N \leqslant k < 2N$$

for $2N$ samples. The procedure for obtaining the correlation function is then to form $n_1'(k)$ and $n_2'(k)$ and therefore $N_1'(j)$, $N_2'(j)$ via the FFT. The product $R'(j)$ is obtained $[R'(j) = N_1'(j), N_2'(j)]$, and then $r'(k)$ via an inverse FFT. This procedure is discussed more fully in the paper by Bergland and the references therein.

As explained in Chapter 1, we can compute the power spectrum of the signal from the Fourier transform of the correlation coefficient or directly from the Fourier coefficients $N(k)$ of the sampled waveform via

$$P(k) = N(k) \cdot N^*(k)$$

6.16.6 Digitizing images

In this section, we shall look at the digitization of images received from a camera. Such a camera may be based on a vidicon, CCD, or a CID as discussed in Chapter 3, and the output of the camera will be a serial signal whose detailed relationship to the image depends on the readout mechanism of the device. For example, in a vidicon system, where the target is raster scanned by an electron beam, the serial signal will correspond to the series of horizontal scan lines used to read out the image. The analog variation of the signal intensity corresponds to the intensity of the image along that scan line. With CCDs and SAiPDAs, the serial signal is a train of pulses, each pulse corresponding to one picture element, or pixel, of the device.

The next stage in this process is to convert this analog signal into a digital one, and then to store the information in some form of memory. In high quality systems, where fast scene changes may occur, a 'frame grabbing' technique is employed. In this system, one frame of the image is sampled, digitized, and stored during the interval between frames. An ADC assigns a digital level to each analog signal level and such converters usually quantize into 8, 64, or 256 levels. The image is then said to be quantized into a number of *gray* levels. The gray scale quantization used depends on the image quality desired. For example, the eye is capable of distinguishing about 15 shades of gray but is more sensitive to the difference in brightness of adjacent gray shades. The sampling rate of the ADC must be at least twice as fast as the highest frequency components present in the signal, which usually means sampling at megahertz frequencies. For example, consider digitizing TV signals in the UK. Such signals have a bandwidth of 5.5 MHz so that the sampling rate is 11 MHz at least. In practice, the sampling rate is higher than this and, for TV, is taken as an integral multiple of the color subcarrier frequency (4.43 MHz). Thus, a TV format digitizer would operate at sample rates of 17.72 MHz. It is obvious from this discussion that a flash ADC is usually employed in such systems.

In situations where the experiments can assume that the object under view will remain fairly static, another technique, employing slower and cheaper circuitry, can be employed. In this case, the microprocessor interface assembles a single image form pieces obtained from many frames. Between frames, the ADC converts a small portion of the image, at successively later times during the frame, storing the digitized information. Eventually, a complete image is formed. The ADC samples typically at 100 kHz to 1 MHz. Of course, the success of the technique depends on all the frames used to build an image being the same.

The digitized image is then usually stored in semiconductor memory prior to long term storage. The amount of such semiconductor memory needed can be prohibitive if we store the digital value of every pixel in the image (direct bit-map image store). For example, a standard format would be 512×512 pixels and using only two gray levels would require $2 \times 512 \times 512 = 262\,144$ bits. This may be a problem for microprocessor-based systems and data compression techniques are often employed. We may, for example, store binary images (two gray levels) as the lengths of strings of identical pixels (5 black, 7 white, 2 black and so on; see Pratt 1978).

An example of a microprocessor-based image processing system has been discussed by Cady and Hodgson (1980). The camera employs a 100×100 pixel CCD light sensor having a 3 mm by 4 mm area. A block diagram of

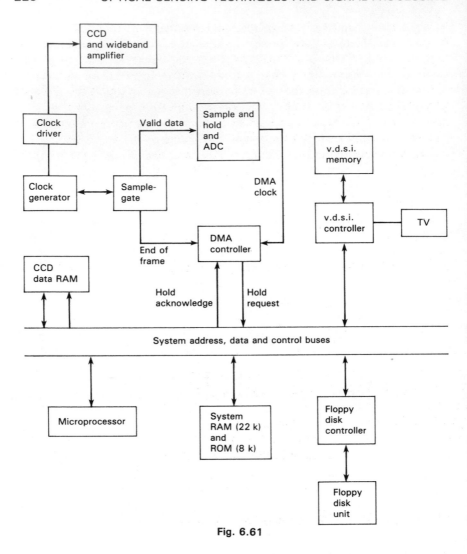

Fig. 6.61

the system is shown in Fig. 6.61. The clock generator produces the required clocking signal for CCD readout and also provides a signal which indicates when valid data is available. This signal therefore indicates to the sample and hold circuit when to sample the signal and starts the analog to digital conversion. The output from the camera is a serial train of analog pulses, each one corresponding to the output of one pixel. These pulses are converted to an eight-bit word. The whole system is controlled by an eight-bit microcomputer (Intel SBC 80, employing an Intel 8080 microprocessor) and this generates a signal which is sent to the CCD camera to begin the capture of the next frame of data. The 1 MHz data rate dictated by the CCD readout requirements means that DMA techniques must be used to transfer data to main memory subsequent to long term data storage or data display. In order to display the current or any previously stored image on a TV or video monitor, the system has a visual display store interrogator (v.d.s.i.) which can read out data stored in a fast dynamic RAM (v.d.s.i. refresh memory) and convert it to a TV-compatible video signal. The v.d.s.i. RAM is large enough to store two complete frames of data and its contents may be changed by the microprocessor. The microcomputer contains software which divides into two sections. The first is operating system software which provides general 'housekeeping' functions such a starting frame capture, storing images on disk, and so on. The second section contains image processing software.

Digital image processing starts with the image of an object and produces a modified version of that image. We may consider two separate processes – image restoration and image enhancement. Image restoration attempts to remove the effect of some degradation process on the image. There are many possible sources of such degradation; for example, saturation of the detector, motion blurring of the image, diffraction limited optical systems, and so on. Restoration techniques require some knowledge of the degradation if an attempt is to be made to reverse its effect. Image enhancement is an attempt to improve the appearance of an image for human viewing or subsequent machine processing (Andrews 1974).

If we have some image $d(x, y)$ of an object $s(\alpha, \beta)$ which is degraded by some operator $h(x, y, \alpha, \beta)$ then

$$d(x, y) = \int_{-\infty}^{+\infty} \int_{-\infty}^{+\infty} s(\alpha, \beta) h(x, y, \alpha, \beta) \, d\alpha \, d\beta + N(x, y)$$

The operator $h(x, y, \alpha, \beta)$ is known as the point spread function. $N(x, y)$ represents the noise. Usually, it is assumed that the point spread function is spatially invariant and, in this case,

$$d(x, y) = \int_{-\infty}^{+\infty} \int_{-\infty}^{+\infty} s(\alpha, \beta) h(x - \alpha, y - \beta) \, d\alpha \, d\beta + N(x, y)$$

The form of the point spread function depends on the degradation source. These may be point degradations, spatial degradations, temporal degradations or chromatic degradations. In order that a digital image restoration system may be designed effectively, it is necessary to quantify the image degradation effects of the imaging system, digitizer, and display. Producing a model of the degradation allows us to produce a computer algorithm to invert the model algorithm. There are two basic approaches to the modeling of image degradation effects – *a priori* and *a posteriori* modeling. *A priori* modeling is used when information is available beforehand concerning the degradation. For example, it may be known that certain geometrical relationships exist between object and image during exposure, bringing about coordinate blur which may be corrected for in software. *A posteriori* modeling uses the actual image as an aid to determining the parameters describing the degradation. For example, knowledge that an edge or point exists in an image may be used to model the point spread function.

Traditional one-dimensional signal processing techniques have been used successfully in two dimensions for image restoration. For a linear spatially invariant point spread function with additive noise, restoration can be performed with linear filtering techniques. Such techniques are employed in one of the most common image restoration tasks, that of spatial image restoration to remove geometric distortion, compensate for image blur, and diminish noise effects. In the absence of noise, we can use Fourier techniques and write

$$D(u, v) = H(u, v)S(u, v)$$

where D, H, S are the Fourier transforms of d, h, s respectively. This equation implies that we can determine $S(u, v)$ by multiplying $D(u, v)$ by the inverse of $H(u, v)$; that is, we apply the inverse filter

$$R(u, v) = [H(u, v)]^{-1}$$

However, such a filter may not exist and furthermore the noise has been ignored. Taking these factors into account, several filters have been proposed, such as the Wiener filter

$$R(u, v) = \frac{H^*(u, v)}{|H(u, v)|^2 + \dfrac{\phi_n(u, v)}{\phi_f(u, v)}}$$

where ϕ_n and ϕ_f are the noise and object power spectra respectively. Several variations of the Wiener filter have been developed for image restoration, for example the geometric mean filter

$$R(u, v) = \left\{\frac{1}{H(u, v)}\right\}^s \left\{\frac{H^*(u, v)}{|H(u, v)|^2 + \dfrac{\phi_n(u, v)}{\phi_f(u, v)}}\right\}^{1-s}$$

where $0 \leqslant s \leqslant 1$ (see Pratt 1978).

Image enhancement seeks to improve the visual appearance of an image or to convert the image to a form better suited to human or machine analysis. For example, an image enhancement system may emphasize the edge outline of an image prior to, say, computing the size and shape of an object. Several enhancement techniques are available (Andrews 1974). Intensity manipulation involves an operation on the image $d(x, y)$

$$d(x, y) = I[d(x, y)]$$

where I is a nonlinear mapping of d independent of the position (x, y) in the image. A prime example of this is the correction of sensor nonlinearities. Eye modeling involves an understanding of the psychophysical aspects of human vision to allow the computer to enhance an image by precompensating for the visual system. Edge sharpening is often implemented in the frequency domain of an image using filters which have high frequency bandpass characteristics. Finally, we may enhance an image by using color on an originally monochrome image. It may be employed to make a human observer pay more attention to certain objects than if they were colored normally, or to match the color sensitivity of the human eye.

SUMMARY

In this chapter we have looked at the area of digital signal processing. The first stage in this process is to digitize the analog signal. This involves sampling the signal at least at twice the highest frequency present, quantizing the signal, and finally assigning a digital code to the quantized level. The integrated circuits, the analog to digital converters (ADCs), the digital to analog converters (DACs) used to digitize the analog signals were discussed in detail. Several types of data converter were mentioned, with emphasis placed on their method of operation. The errors associated with data converters were next discussed in detail, and this allowed a determination of the relative accuracy of a specified converter to be made. The maximum frequency that can be digitized by a converter was shown to be remarkably low (of the order of a hundred hertz for a 10-bit converter with a conversion time of a few microseconds) and the use of a sample and hold circuit to increase this frequency was discussed. A complete data acquisition system could now be developed and two approaches were introduced and compared.

The use of microprocessors in this area was inevitably introduced. The basic elements of a microprocessor-based computer (or microcomputer) were discussed with special emphasis being placed on the way in which a microcomputer communicates with the outside world. Synchronization of data transfer between the microcomputer and an external device, such as a data converter, is obviously important and the concept of 'handshaking' was introduced. The hardware aspects of data transfer were discussed, with both parallel and serial data transfer appearing as valid means of data flow. Rather than construct such input/output hardware interfaces from discrete integrated circuits, large scale parallel (PPIs) or serial (UARTs) integrated circuits may be used and a circuit was developed to show how a microcomputer possessing a PPI could be interfaced to an ADC. The concept of bus interface connections was also discussed, in particular the IEEE-488 instrumentation interface bus.

Finally, several examples of the use of microcomputers in data manipulation were provided. These examples included digital signal averaging, digital phase sensitive detection, digital filtering, fast Fourier transforms (FFTs), correlation techniques, and, finally, digitization of images.

REFERENCES

Andrews, H.C. 1974. Digital image restoration: a survey. *IEEE Computer*, May, pp. 36–45.

Bergland, G.D. 1969. A guided tour of the fast Fourier transform. *IEEE Spectrum*, July, pp. 41–52.

Cady, F.M. and Hodgson, R.M. 1980. Microprocessor based interactive image processing system. *Proceedings of the IEE*, Vol. 127, pp. 197–201.

Cheetham, B.M.G. and Hughes, P. 1982. Digital filter design. *Wireless World*, May, pp. 52–4.

Cochran, W.T., Cooley, J.W., Favin, D.L., Helms, H.D., Kaenel, R.D., Lang, W.W., Maling G.G. Jr., Nelson, D.E., Rader, C.M. and Welch, P.D. 1967. What is the fast Fourier transform? *Proceedings of the IEEE*, Vol. 55, pp. 1664–77.

Cooley, J.W. and Tukey, J.W. 1965. An algorithm for the machine calculation of complex Fourier series. *Mathematics of Computing* Vol. 19, pp. 297–301.

Docchio, F., Longeni, A. and Fusi, G. 1980. Microprocessor based systems for optical spectroscopy. *Euromicro Journal*, Vol. 6, pp. 316–20.

Heffer, D.E., King, G.A. and Keith, D. 1981. *Basic Principles and Practice of Microprocessors*. London: Edward Arnold.

Higgins, R.J. 1976. Fast Fourier transform: an introduction with some minicomputer experiments. *American Journal of Physics*, Vol. 44, pp. 766–73.

Iacopini, E., Smith, B., Stefanini, G. and Carusotto, S. 1983. Digital techniques applied to phase sensitive detection. *Journal of Physics E: Scientific Instruments*, Vol. 16. pp. 844–7.

Lam, R.B., Sparks, D.T. and Isenhour, T.L. 1982. Cross correlation signal to noise enhancement with applications to quantitative gas chromatography/Fourier transform infra-red spectroscopy. *Analytical Chemistry*, Vol. 54, pp. 1927–31.

Larson, T. and Dyrik, G. 1985. Fast Fourier transforms using a microcomputer. *Wireless World*, September, pp. 80–82.

Lynn, P.A. 1973. *An Introduction to the Analysis and Processing of Signals*. London: Macmillan.

Momo, F., Ranieri, G.A., Sotgui, A. and Terenzi, M. 1981. Microcomputer based phase sensitive detector. *Journal of Physics E: Scientific Instruments*, Vol. 14, pp. 1253–6.

Nagel, H.T. Jr. and Nelson, V.P. 1981. Digital filter implementation on a 16-bit microcomputer. *IEEE Micro*, February, pp. 23–41.

Oppenheim, A.V. and Weinstein, C.J. 1972. Effects of finite register length in digital filtering and the fast Fourier transform. *Proceedings of the IEEE*, Vol. 60, pp. 957–76.

Pratt, W.K. 1978. *Digital Image Processing*. New York: John Wiley.

Ricci, D.W. and Nelson, G.E. 1974. Standard instrument interface simplifies system design. *Electronics*, November, pp. 95–106.

Saniie, W.J. and Luukaala, M. 1983. Digital phase sensitive detection based on inphase and quadrature sampling. *Journal of Physics E: Scientific Instruments*, Vol. 16, pp. 844–7.

Shannon, C.E. 1948. A mathematical theory of communication. *Bell Systems Technical Journal*, Vol. 27, pp. 379–423 and 623–56.

Singleton, R.C. 1969. Algorithm 339, an Algol convolution procedure based on the fast Fourier transform. *Communications of the Association of Computing Machinery*, Vol. 12.

Tan, B.S. and Hawkins, G.J. 1981. Speed optimized microprocessor implementation of a digital filter. *Proceedings of the IEE*, Vol. 128, pp. 85–93.

PROBLEMS

6.1 Calculate the maximum frequency that can be digitized by (a) an 8-bit, and (b) a 10-bit ADC having a conversion time of 10 μs. If a sample and hold circuit having an aperture time of 100 ns is placed before the ADC, calculate the new value of the maximum frequency.

6.2 A TV picture is to be digitally encoded. It has a resolution of 625 lines by 400 pixels per frame and there are 25 frames per second. Determine the minimum sampling rate of the digitizing ADC.

6.3 A 12-bit ADC has the following errors:

Linearity error	$\pm \frac{1}{2}$ LSB
Temperature coefficient	5 p.p.m./$^\circ$C
Power supply drift	0.05% for a 1% change in power supply.

If the ADC has to work over a temperature range of 50°C and the power supply exhibits a 5 percent drift, calculate the relative accuracy of the ADC.

6.4 A 32-channel analog switch is used as a multiplexer in a data acquisition system. The 'on' resistance of each channel is 1.5 kΩ and each 'off' channel provides 5 nA of leakage current into 50 pF of capacitance. Determine: (a) the offset voltage of the device; (b) the settling time of the switch. If the ADC in the system is a 12-bit ADC with a conversion time of 10 μs, determine in what ways the analog switch limits the operation of the data acquisition system.

6.5 Using a digital to analog converter, a comparator, and a microcomputer with a PPI, design an analog to digital converter. Sketch the software required using both polling and interrupt-driven handshaking.

6.6 A digital filter has the recurrence formula

$$y(n) = -\alpha y(n-1) + x(n) - x(n-1)$$

Determine the type of filter this represents and sketch its transfer function.

6.7 Starting from a simple analog RC high pass filter, develop a recurrence formula for this filter. Hence, implement the high pass digital filter in hardware and in software, using a high level language such as BASIC.

Appendix

INTEGRATED CIRCUIT SEMICONDUCTOR TECHNOLOGY

A typical data acquisition board will contain a number of integrated circuits, ranging from large scale integrated (LSI) circuits such as microprocessors and memories, through medium scale integrated (MSI) devices such as counters, down to small scale integrated (SSI) devices such as gates, in addition to analog chips such as op-amps. In most situations, SSI and MSI devices are made using the bipolar transistor technology known as TTL whilst LSI devices use a field effect transistor technology known as MOS technology. Analog devices are chiefly made from bipolar transistors although there is an increasing trend to introduce MOS devices or chips which mix the two technologies. In this section, we will discuss the major characteristics of the different semiconductor technologies employed in integrated circuits. Some of the significant characteristics we might look for in a technology are:

1. Speed, or how quickly a gate can switch between two logic levels. This factor is usually quoted as the propagation delay of the gate.
2. Density, or how many gates can be packed into unit area. This factor will depend on the power dissipated by the gates.
3. Noise immunity or the range of supply voltage and currents over which the technology operates properly.
4. TTL compatability, an important factor since most electronic systems are built with standard TTL circuits.
5. Cost and maturity of the technology.

Such information is of use to the designer who will need to decide on the technology to employ for a specific application by trading off the characteristics of the different technologies. For example, a high speed technology will have its logic levels close together and so will have a low noise immunity.

Bipolar technologies

There are three major technologies based on the bipolar transistor. These are transistor-transistor logic (TTL), emitter coupled logic (ECL), and integrated injection logic (I^2L). In general, bipolar technologies are faster than MOS technologies, are capable of operating at much higher frequencies, and can deliver much higher currents. On the deficit side, bipolar technologies have higher power dissipation than MOS, do not have the packing density (smaller number of transistors in a given chip area) and have lower noise immunity.

The most widely used family of logic circuits is TTL. SSI and MSI TTL devices are usually distinguished by a numerical designation 74xxx (where xxx is a two or three digit designation which determines the type of device, e.g. 7404 is an inverter). These have typical switching speeds of 35 mHz. Schottky TTL devices (74 Sxxx) have much higher speeds than the standard devices, (typically 125 MHz) while low power Schottky TTL (74LSxxx) offer the additional advantage of low power consumption. This latter family is one of the most popular TTL families at present. Recently there has been the introduction of the so-called FAST series of TTL devices (74Fxxx) which stands for Fast Advanced Schottky TTL which have a 30 percent improvement in speed over Schottky TTL while consuming only one quarter of the power.

Emitter coupled logic is the standard technology for very high frequency operation (up to 1500 Mhz in some cases) although the advent of chips based on gallium arsenide will be important in this area in the future. ECL has a much higher power consumption than TTL, has a lower noise margin, uses twin power supplies and operates at different voltage levels from TTL, so that interfacing with TTL requires extra circuitry. ECL is probably only a worthwhile technology to use when speed is of the ultimate importance.

Integrated injection logic is the least mature of the three bipolar technologies. It has a very low power consumption and can operate at moderately high frequencies with TTL switching speeds. It is one of the simplest semiconductor technologies and allows both digital and analog circuitry to be placed on one chip, especially useful in such chips as ADCs.

MOS technologies

MOS devices are produced as NMOS (n-channel) or PMOS (p-channel) devices, or a combination of these two, CMOS (complementary MOS) devices. Most microprocessors and memory chips are made from MOS technology, whilst in the SSI and MSI area the traditional dominance of TTL is now being challenged by recent advances in CMOS technology. MOS devices have very high densities, low power dissipation and have good noise immunity. CMOS devices in particular have exceptionally low power consumption making them ideal for use in battery powered devices. However, they are generally slower than bipolar devices, have low output currents and generally work at different voltage levels from TTL.

PMOS devices were the first MOS devices but are now relatively rare. NMOS devices are denser and faster than PMOS and are now generally TTL compatible. Most popular microprocessors and memories are NMOS devices. CMOS devices are based on the use of n-channel and p-channel devices connected in parallel. Conduction only occurs when the device is actually switching from one state to the other, resulting in very low power consumption (although this will increase at high operating frequencies). CMOS devices can also operate over a wide range of power supply voltages (from 3V to 20V). Recent advances in CMOS technology have resulted in much increased speed of CMOS devices. Most microprocessors and memories are available in CMOS, and SSI and MSI CMOS chips are now challenging TTL. The early series of CMOS MSI and SSI chips (the 4xxx) series were very slow (working up to a maximum of about 5 MHz). Recently, a new family of CMOS MSI and SSI chips have appeared having the speed of TTL while retaining the low power advantage. These are the high speed CMOS (74HCxxx) and generally have the same pinout as the corresponding 74xxx series of TTL chips. These devices also have the same current drive as TTL, the high noise immunity of CMOS but operate over a more restricted power supply range than standard CMOS (2 V to 6 V). A further variation on this theme is the HCT version of high speed CMOS. These have input switching levels that are compatible with 74LSxxx series and operate with power supplies between 4.5 V and 5.5 V.

THE COMPARATOR

The comparator is a circuit which is properly discussed in a chapter on analog systems, but it is a circuit which is most frequently used in digital systems. The comparator provides an indication of the relative value of two voltages. If one of these voltages is some reference voltage, then the comparator will indicate whether the other voltage (the signal voltage) is above or below the reference voltage.

The comparator is basically an op-amp which has no feedback connections and is illustrated in Fig. 1.

In this configuration, the op-amp has a gain of around 100,000. Let us suppose that $V_{in} - V_{ref} = 100 \ \mu V$, then the output voltage would be around $+10$ V. Similarly, if $V_{in} - V_{ref} = -100 \ \mu V$, then the output voltage would be about -10 V. It is obvious from this discussion that, for very small differences in the differential voltage at the input of the op-amp, the output voltage will rapidly approach the supply voltage (either the positive or negative value) beyond which, of course, it cannot go. The output voltage is said to saturate (at a value usually a volt or two below the supply). Thus, when the input voltage is slightly less than the reference, the output voltage is at the negative saturation voltage, $-V_{sat}$, while if the input voltage changes to a value slightly above the reference, the output of the comparator will switch to $+V_{sat}$. A plot of the output voltage as a function of the differential input voltage to the comparator is shown in Fig. 2.

It should be noted that, since the voltage difference needed to switch the comparator from one state to the other is very small, the dominating factor that determines the exact switching level is the offset voltage of the amplifier which should be nulled out.

The comparator is quite often used with digital electronics where voltages other than the saturation voltages of the op-amps are used. In order to interface to these devices, the output voltage of the comparator must be clamped in some way. Figure 3 shows how this can be done to allow the comparator to interface with so-called TTL integrated circuits where voltages between 0 V and 5 V are employed.

When the input signal to the comparator is a slowly varying one, the output will change slowly which may be a disadvantage when fast switching is needed such as in digital circuits. To overcome this, positive feedback is employed, as illustrated in Fig. 4.

Let us suppose that the output voltage is positive. The actual voltage at the $+$ input to the op-amp is

$$V_+ = V_{ref} + [V_{sat} - V_{ref}](R_1/R_1 + R_2)$$

As the input voltage begins to exceed this value, the output of the comparator begins to fall towards $-V_{sat}$. However, this falling voltage is also fed back into the $+$ input of the op-amp and this regenerative action will cause the comparator to switch rapidly into the opposite state. The voltage at the $+$ terminal of the op-amp is now

$$V_- = V_{ref} + [-V_{sat} - V_{ref}](R_1/R_1 + R_2)$$

There are now two switching voltages and we have now introduced hysteresis into the comparator $V_+ - V_-$.

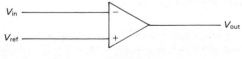

Fig. 1

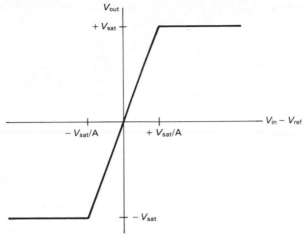

Fig. 2

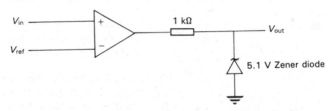

Fig. 3

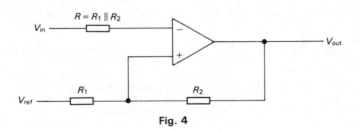

Fig. 4

As well as increasing the switching speed of the comparator, hysteresis is useful when there is low-level noise on the input voltage. Without feedback, this noise would cause the comparator to switch back and forth rapidly. By making the hysteresis larger than the noise, this effect can be eliminated.

DIGITAL SIGNAL PROCESSING CHIPS

Although digital signal processing can be accomplished using microprocessor techniques, there is an increasing trend to produce integrated circuits specifically for the purpose. Digital signal processing involves, as its most common operations,

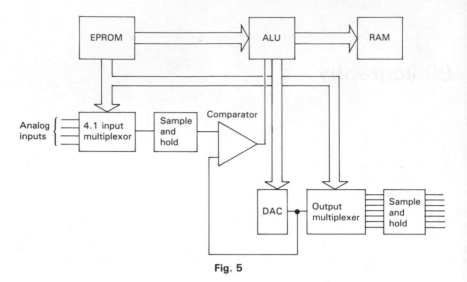

Fig. 5

multiplication and addition, and the storage and retrieval of data. Advances in integrated circuit technology have enabled some or all of these functions to be accommodated on a single chip. One of the first such custom digital signal processors was the Intel 2920 analog signal processor.

This single-chip device takes in analog signals, converts them to digital form where they can be processed by a dedicated on-chip microprocessor, and finally converted back to analog form for output. The analog signals are fed into the system via a four-channel input multiplexer where they can be sampled by a sample and hold converter and converted to digital form by a nine bit successive approximation ADC (in fact made up of a 9 bit DAC and a fast comparator). The microprocessor section consists of an EPROM for storage of instructions, some RAM and an ALU. The RAM is organized as 40 words, each of 25 bits, and data is processed in the ALU using 25-bit two's complement arithmetic. The total system is controlled by the program in the EPROM and the sampling rate is determined by the length of the program. All instructions have a fixed execution time (600 ns or 800 ns depending on the device). An 800 ns device contains 50 instructions and has a sampling period of 40 μs, allowing a sampling rate of 2500 samples per second. The 2920 has eight analog output channels, each one having a sample and hold circuit demultiplexed from a common 9 bit DAC. A simplified block diagram of the 2920 is shown in Fig. 5.

Other manufacturers produce digital signal processing chips but usually without the data converters. An example of this is the Texas Instruments TMS 320 single chip microcomputer. It can perform 32-bit arithmetic and will, using on-chip parallel multipliers, typically carry out 16×16 bit multiplication to produce a 32-bit result in 400 ns. The device runs at 20 MHz and 90% of its instructions are executed in 20 ns. The microcomputer contains both ROM and RAM and has facilities for expansion of these.

Bibliography

Beauchamp, K. and Yen, C. 1980. *Data Acquisition for Signal Analysis*. London: George Allen & Unwin.

Beeforth, T.H. and Goldsmid, H.J. 1970. *Physics of Solid State Devices*. London: Pion Ltd.

Bibbero, R.J. 1977. *Microprocessors in Instruments and Control*. New York: John Wiley.

Carrick, A. 1979. *Computers and Instrumentation*. London: Heyden.

Castleman, K.R. 1979. *Digital Image Processing*. Englewood Cliffs, NJ: Prentice-Hall.

Connor, F.R. 1973. *Noise*. London: Edward Arnold.

Coughlin, R.F. and Driscoll, F.F 1982. *Operational Amplifiers and Linear Integrated Circuits*. Englewood Cliffs, NJ: Prentice-Hall.

Garrett, P.H. 1978. *Analog Systems for Microprocessors and Minicomputers*. Reston, Va.: Reston.

Gowar, J. 1984. *Optical Communication Systems*. Hemel Hempstead: Prentice-Hall International.

Hamming, R.W. 1983. *Digital Filters*. Englewood Cliffs, NJ: Prentice-Hall.

Intersil 1980. *Data Acquisition Handbook*. Intersil Inc.

Kittel, C. 1980. *Introduction to Solid State Physics*. New York: John Wiley.

Lynn, P.A. 1980. *An Introduction to the Analysis and Processing of Signals*. London: Macmillan.

Motchenbacher, C.D. and Fitchen, F.C. 1973. *Low Noise Electronic Design*. New York: John Wiley.

Ott, H.W. 1976. *Noise Reduction Techniques in Electronic Systems*. New York: John Wiley.

Rose, A. 1963. *Concepts in Photoconductivity*. New York: John Wiley.

Ross, D.A. 1979. *Optoelectronic Devices and Optical Imaging Techniques*. London: Macmillan.

de Sa, A. 1981. *Principles of Electronic Instrumentation*. London: Edward Arnold.

Smith, R.A. 1959. *Semiconductors*. Cambridge: Cambridge University Press.

Stone, H.S. 1982. *Microcomputer Interfacing*. London: Addison-Wesley.

Sze, S.M. 1969. *Physics of Semiconductor Devices*. New York: John Wiley.

Titus, J.A., Titus, C.A., Rony P.A., and Larsen, D.G. 1979. *Microcomputer–Analog Converter*. Indiana: Howard W. Sams.

van der Ziel, A. 1956. *Noise*. London: Chapman & Hall.

Wilson, J. and Hawkes, J.F.B. 1983. *Optoelectronics: an introduction*. Hemel Hempstead: Prentice-Hall International.

Yariv, A. 1985. *Optical Electronics*. Eastbourne: Holt Saunders.

Zaks R. and Lesea, A. 1977. *Microprocessor Interfacing Techniques*. Sybex Inc.

Index